Thrive

Thrive

Practical Strategies to Nourish Teacher Wellbeing

DANIELA FALECKI

amba
press

Published by Amba Press
Melbourne, Australia
www.ambapress.com.au

Cover designer: Tess McCabe

ISBN: 9781923116146 (pbk)
ISBN: 9781923116153 (ebk)

A catalogue record for this book is available from the National Library of Australia.

Contents

Preface

Sitting at my desk with my head in my hands, I tried to hide the tears as they rolled down my face. An argument with a student, an abusive email from a parent and a deadline looming over me. I was done. The tank was empty. The system had won.

Fifteen years of full-time teaching and the job was only getting harder. At first I'd thought I simply needed to work smarter, stay longer, be better organised and learn more. After continuously running on unreasonable timelines and drowning in administration, I had nothing left to give.

I had always wanted to teach. Maybe it was because I was the eldest of four children in a dysfunctional family where responsibility lay heavily on my shoulders. Maybe it was just because I liked bossing people around. Either way, I was an excited, passionate and giving teacher. I wanted to make a difference in the world by helping people believe in themselves, and I loved supporting them to thrive.

As I began my career, there were early signs of the disparity between what I wanted to do—help young people grow and learn—and the reality of ticking boxes to meet outcomes and ensure compliance. Nothing prepares you for your first year of teaching. University prepares you with theory, but not for the dynamic reality of a school. It's a bit like enrolling in a driving course but not getting to handle a car. During my first year of teaching, I felt like I was trying to merge lanes on a busy highway in peak hour with everyone and everything moving fast around me. Demands were high and resources low. You had to decide very quickly whether you were willing to continue on your chosen path.

I was given the honour of organising whole-school sport every week—apparently a traditional way of welcoming new Phys Ed teachers! This was not something I had done before or been trained to do. In addition to my already overwhelming teaching duties, I was responsible for the movements of 900 students and 70 staff for three hours every Tuesday afternoon. This was event management on a large scale, from arranging try-outs to deciding draws, booking buses and allocating teams to venues. I developed systems for collecting records of attendance, scores, incidents and more. I busily created staff information sheets, instruction folders and student manuals with tips relating to uniform and expectations. In true perfectionist style, I colour-coded and laminated all folders with bold headings and clear instructions.

But too often it seemed that my efforts were in vain. Every week I was left shaking my head, feeling like I was herding cats. It wasn't the students: my fellow teachers would get on the wrong bus or neglect to return folders. Why weren't my colleagues working with me? Why were they making my job so much harder than it needed to be? Why was no-one on my side?

I later came to realise that everyone was doing the best they could with the resources they had. I have since learned that many teachers are struggling, and that we all respond in different ways. What I didn't realise at the time was that I was my own worst enemy.

Getting in my own way

Looking back, I cringe at my induction and am disappointed in the way I handled my stress. I put a lot of pressure on myself to 'be right' and 'get it right'. If things went wrong, I would push myself to work longer and harder. I took it personally if someone gave me negative feedback or highlighted something I had missed. I armoured my vulnerability with defensiveness, thinking I was being strong and professional when in fact I lacked social and emotional skills. Each day I would wake up determined to give more and do more, only to find myself becoming completely exhausted. I didn't know how to listen to my body, and I certainly didn't know how to communicate my overwhelm. If I wasn't working, I was feeling guilty about it. My inner critic was strong and it drove my behaviour. To my colleagues I was a high-functioning go-getter, but most days I wanted to curl up in a corner and cry.

I was chasing a feeling of satisfaction, of knowing that what I had done was enough. The problem was that nothing ever seemed enough. 'If I can get this finished, I will feel satisfied,' I thought. Just as one task was completed, another three filled the space. No matter what I did, the job was never completely done. It seemed that the harder I worked the more there was to do.

I thought being a good teacher meant doing a lot of work and doing it well. I thought you had to sacrifice yourself and give everything to your students. There was no time for whinging. You had to be creative, work with what you had and get the job done regardless of circumstance. I didn't realise there was another way. I didn't realise I was running to a place that had no finish line. I didn't realise I was allowed to enjoy the journey and in fact celebrate my hard work and impact. I didn't have the skills or resources to even think about my own health and happiness. Conversations about teacher wellbeing didn't exist.

I needed support, but I didn't ask for it. I thought I could do it all on my own. I didn't have anyone by my side because I wouldn't allow it. Teachers seem to have an unspoken martyr archetype where we put ourselves last—we use the word 'busy' as a trophy. Persisting with my invisible martyr badge for years, I followed this norm perfectly. After blaming myself for not being good enough, I expanded the blame to the other teachers, the school leaders and the 'system'. I had moved from martyr to victim. The answer, or so I always thought, was simple: to change schools. The next one would be better.

My next school may have had different faces, but it had the same pressures. The job was still tough with constant curriculum changes, and compliance and expectations continued to overwhelm me. With each advancement in technology came more to do. Student needs seemed to grow, as did the drive for parental involvement. That might have made a difference if the parents who actually came to interviews were those with the more demanding children.

The strong sense of responsibility that had been drummed into me as an eldest daughter became a double-edged sword for me as a teacher. I felt responsible for everyone in my care. I tried to fix people and situations. I put immense pressure on myself to help those who needed me. After all, I had entered teaching because I deeply cared about people. I knew I could be of service to them, and went above and beyond to do just that. I listened at lunchtime, created fancy resources and advocated in welfare meetings. I diversified lessons for gifted students and offered extra lessons for others. In the holidays I would research new methods. I loved helping people, but it came at a high cost. I didn't know how to rest. I chose not to switch off, and no-one told me about the importance of emotional and mental recovery. Here I was putting all my energy into others when I needed to put energy into myself.

Cynicism and the turning point

Through sheer perseverance, I had now been a government high school teacher for 10 years. I still loved my job, insisted on catering to everyone's needs, was creative in program development and worked long hours. But I had also become *that* teacher. The one who rolls their eyes at new initiatives, who constantly argues and complains about decisions. Yes, I was now the cliché jaded teacher huffing and puffing her way through the day. I had swapped my martyr and victim badges for cynicism, and I liked it. I decided it was easier to care less and do the bare minimum of what was required. Don't get me wrong, I still loved my students and created engaging lessons. But when it came to administration requirements, meetings or staff events, I participated as little as possible. In my mind there was no collective goal we were working towards. Everyone was doing their own thing in their own way with few opportunities for collaboration or shared learning. As long as I turned up on playground duty, I was left to my own devices. The problem was that I wasn't happy.

I started to question my life. Was I really going to keep doing this for the next 40 years? There had to be a better way. I couldn't change the system, so I had to change myself. I made the brave decision to seek counselling and embark on a journey of self-discovery. I was curious about new ways of thinking and being. After all, my way simply wasn't working.

In classic control-freak style, I went to my first counselling session with a list of what I thought to be the key issues. I told my therapist I needed three sessions before I would be 'fixed' and on my way. She sat patiently with a compassionate smile and suggested we get started. Two years later, and after weekly visits, I walked into my last session without a list, without an agenda and open to possibility. Surprisingly, I was now comfortable with aspects of the world I couldn't control because I now had the skills to navigate change as it happened.

I had learned how to listen to my body, how to read emotions as signposts and how to practice self-care. I understood how I processed stress and ways to manage it. I knew how to speak up and ask for help when I needed it, and how to appropriately communicate my feelings. I finally recognised that no-one could 'make me' feel valued or 'make me' work less. I was building my social and emotional competence. I had been teaching these skills to students for years as a health educator, yet here I was applying them to myself for the first time.

My eyes were opened to a different way of being. I took ownership of my decisions and learned to forgive myself when things didn't work out perfectly. I learned how to better communicate, how to be vulnerable and how to build psychological safety in relationships. A new chapter was emerging in my interactions with others. I saw through a lens of curiosity instead of perfectionism. I felt centred, confident and competent.

The seeker in me had been awakened. It was time to merge my personal growth with the teaching world. My journey of learning psychology, exploring new sciences and finding meaningful ways of thinking and behaving had begun. I left the security of the public system and threw myself into Rudolf Steiner education. This would be the answer I was looking for.

There's no magic answer

Reality hit with a bang. The philosophy of Rudolf Steiner education centres on inclusion with holistic learning principles. This includes an awareness of psychological influences on behaviour, and an acknowledgment of the emotional states of students and teachers. The more I learned about the philosophy, the more excited I became. Being able to teach in a Steiner school would meet all my needs. I applied for a job and got it. Unfortunately, the pressures in this school were the same as those everywhere else: high demands, low resources and poor role clarity. If anything, the demands in this small school were greater with fewer people to help spread the load. I would start at 7am and leave at 7pm, four days a week. This didn't include the weekly parent-teacher meetings and weekend stalls you were expected to attend. It got to the point where I was driving to work in tears of exhaustion, finding a way to function through the day then crying again on the way home because there was still so much more to do. I couldn't sustain it any longer.

I made the difficult decision to walk away from the classroom after 15 years. I had learned so much, and I knew there was still more to learn. I needed to leave in order to heal and restore.

I am a strong advocate for Steiner education and its philosophy. But I mistakenly kept looking at my environment when what I really needed to do was to look within. No school had the magic answer I was seeking. Schools by their very nature are structured in ways that don't allow for choice or autonomy. I was told what teach, who to teach, when to teach, when to assess, when to eat and when to take holidays. Managing my wellbeing, however, somehow remained my responsibility!

Student wellbeing begins with teacher wellbeing

I may have left the classroom, but I've always remained close. For the past 10 years I have worked with thousands of teachers across early childhood, primary, secondary and tertiary institutions through my company Teacher Wellbeing. The more I move between schools, the more calls for support I hear. Whether its public or private schools, teaching can be challenging. Every new staffroom presents a mirror of my former self. I see low social and emotional competence, a lack of psychological capital and the failure of a system to support its teachers. I listen with compassion and offer small insights where possible. If these teachers only had my new eyes. If they only knew there was another way. This perspective is what led me to my mission: helping teachers thrive.

It fills me with joy to see big-hearted teachers move from exhausted to energised through my workshops, consulting and coaching. Transformative change can happen across long-term projects and one-day events alike. Educators need opportunities to stop, reflect and connect in meaningful ways. I continue to be inspired by their bravery in challenging the status quo. It is an honour to advocate for their needs and to be a voice for so many who don't feel heard.

It is past time to address the wishes of teachers for greater balance in their lives. Teaching isn't about *doing* more; it's about *being* more. It is important to practice self-care, and the system has a responsibility to support you too. We need to see teachers as people. We need to question processes, review policies and work together to give ourselves every opportunity to be our best for our students.

I look forward to sharing in the journey with you as we explore ways in which can thrive.

Always remember... you matter!

Daniela Falecki

Introduction

This book is for you, the teacher. You spend most of your life thinking about other people's children. You put endless time and energy into supporting young people. You inspire, guide, love and contribute to the building of a new generation.

This is not a fluffy self-help book putting pressure on you to be well in a world full of chaos. It's a book to help you navigate that chaos, celebrate the amazing work you do and learn to thrive. It asks you to challenge your thinking, reflect on your feelings and question your actions. To do this you will have to stop and notice what is happening in and around you.

Through my own experience in teaching and working with teachers across the globe, I am excited to share with you these adaptable strategies to support your wellbeing.

Why focus on teacher wellbeing?

Teaching is getting tougher. Globally, teachers have higher work-related stress than other professions (Education Support Partnership, 2021). The stress we experience can cause a decline in physical health, decrease self-confidence and fragment personal relationships (Howard and Johnson, 2004). As well as feeling exhausted, we can experience a sense of powerlessness and isolation that makes us perceive work as meaningless (Howard and Johnson, 2004).

Teaching is an emotional vocation that relies on relationships to foster belonging and connection (Roffey, 2012). Unfortunately, little time is given for teachers to learn and develop their social and emotional competence, which can result in ineffective coping strategies (Jennings, 2012). When we are emotionally exhausted, we may use reactive and punitive responses that contribute to negative classroom climates and student-teacher relationships (Osher et al. 2007). Education researcher John Hattie has linked teacher motivation to student achievement: 'When teachers become burned out, or worn out, their students' achievement outcomes are likely to suffer because [the teachers] are more concerned with their personal survival.' (Hattie, 2013).

Our ability to cope and our potential to flourish are affected by social and emotional stresses from the daily pressures of teaching (Parker et al., 2009). Balancing demands with available resources is emotionally and mentally draining. We struggle to keep up with the relentless pace, to meet expectations and to be our best for students. Our self-worth suffers: with little left to give, we take criticism personally (Parker, 2012). Those in school leadership positions are also faring badly. The 2022 Australian Principal Occupational Health, Safety and Wellbeing Survey identified workload increases, the poor mental health of young people, the demands of parents and micromanagement by policymakers as critical factors that contribute to leader stress (Riley, 2022). This growing evidence of stress and burnout reveals that the wellbeing of educators is in crisis. Research from around the globe continues to highlight teacher stress as a societal concern (CASEL, 2017; Education Support UK, 2021; AIS NSW, 2017).

We must act now. Teachers are the greatest resource in education. The profession's sustainability depends on our wellbeing. So what's the challenge? Wellbeing is complex, but we discuss it as if it's simple.

The complexities of wellbeing

Schools fall into the trap of discussing wellbeing in simple terms, as if multilayered problems can be solved in a 30-minute meeting at lunchtime. We offer yoga after school and put fruit bowls in the staffroom, but does this really help when we are collapsing under heavy workloads?

Wellbeing is strongly influenced by how our internal world interacts with our external world. We must draw on psychology to better understand and develop it. Wellbeing is also personal. Giving people opportunities to think about what wellbeing means to them is very different to having a conversation about stress, overwhelm or exhaustion.

Just as the World Health Organisation (WHO) defines health as 'not merely the absence of disease', wellbeing is not merely the absence of stress. Wellbeing is not something you do at home to be better at work; it is how you *be* at work. We need to develop a shared understanding of what work-related wellbeing means. We need to ask better questions and not assume that we know what others need. We need to give people a common language and opportunities to reflect. Wellbeing is not something you do *to* people. It is something you do *with* people.

What we really want

Ask any teacher what they want and they will say time and energy. They want time because there is simply too much to do, and they want energy because they are feeling tired. When we are asked to do something, the first thing we consider is how much time it will cost us. We then ask ourselves how much energy we will need to give. If we feel we have the time and energy, we feel well. If we feel we don't have the time and energy, we feel unwell. Time and energy become the currencies of our wellbeing.

We are going to unpack these two currencies to explore how we can nourish our energy and take back control of our time. Not so we can do more, but so we can *be* more to the people who matter—including ourselves.

Building your capacity

If we want to move from exhaustion to energy, we need to build our capacity. This means prioritising time for learning. While it would be nice to think that the system will change, we are better placed to focus on what we can control.

I invite you to reflect on your life experiences, work and personal responsibilities to find out what works best for you. The question is 'How can you thrive at work?' This requires time to think and reflective skills to assess choices and options. This is not a book that passes blame on the system, leadership, parents or anyone else. It is a solution-focussed resource to help you be, the best version of you.

Sowing the seeds of wellbeing

At the end of this section, you will understand what wellbeing means in the context of a busy school environment. You will also have a model to help you understand who is responsible for your wellbeing at work and how these layers of responsibility interact with each other.

The wellbeing of individual teachers matters for schools as a whole. When teachers are overwhelmed and stressed, their effectiveness as educators diminishes. This leads to decreased student achievement, disengagement and even behavioural issues.

Teacher wellbeing directly impacts the dynamics of an educational team. Overwhelmed teachers may struggle to collaborate effectively, share best practices or provide support to their colleagues. This lack of cohesion and teamwork can hinder the growth and success of the entire teaching staff.

From an organisational perspective, neglecting teacher wellbeing can have long-term consequences. High turnover rates, absenteeism and decreased job satisfaction are commonly associated with overwhelmed educators. These factors not only disrupt continuity in the classroom but also impose financial burdens on educational institutions.

With an understanding of the complex factors that affect teacher wellbeing, we must turn our focus inwards to better understand the role of stress and how it impacts the body. By identifying signs of stress, we can reduce their impact on our thoughts, feelings and behaviours.

Mindset is a key factor. I have no doubt that most teachers have a harsh inner critic. We will explore how this inner critic can cause us unnecessary stress. When we understand our inner critic, we can engage powerful tools to quieten it. Wellbeing theories can help us build a language that explains what it means to be well, and we can create self-awareness and self-regulation through social and emotional learning.

The basics of teacher wellbeing

Overloaded, overwhelmed, over it

Have you ever tried to explain what it's like to be a teacher? People often remark on the short working hours or the long holidays, and wonder what all the fuss is about when they hear stories of teacher stress.

How do you explain the energy needed to manage ever-increasing student demands? To control your emotions as you repeat yourself for the umpteenth time while secretly wanting to yell, cry or punch someone? To answer a thousand dumb questions? (We say there are no dumb questions, but there absolutely are.)

How do you explain the pressure to push ourselves to meet goals we didn't set? The expectation that we will produce pre-determined outcomes regardless of resources, context or circumstances?

How do you explain the endless hours spent writing policies that no-one will read? The surveys you complete and receive no feedback on? The thousands of hours of professional development you will never use? The parents who blame you for their child's behaviour or the child who tells you that they own you because their family pays your wage?

How do you explain the demands placed upon you with little control over decision-making or resource allocation? The fact that no amount of work is ever enough, that the reward never meets the effort?

You're not a machine

Speak to any teacher and they will tell you the day goes very fast. Time is precious and there is a lot to do. Added to this are constant change and the expectation of learning new things as quickly as possible. Teaching often made me feel like a machine. As the years went on, my to-do list increased alongside my knowledge and competence. I learned how to work faster, but it was never enough. As technology advanced in efficiency so did I, but my proficiency only led to greater expectations. Expanded accessibility meant that the workday never seemed to end. I was judged and rated for the output I produced, expected to function at high speed with little consideration for mental rest or emotional repair.

Machines are typically programmed to take a set of information, manipulate it, and produce an outcome that is repeated over and over. While some form of consistency is necessary for equality and inclusivity in education, it is also important to remember that teachers have their own needs and interests. Each of us has a head that thinks, a heart that feels and an energy system that requires constant renewal. It is this energy system that we engage with each day. We give energy and need energy to renew ourselves. We must stop measuring our effectiveness as educators by what we do, and start connecting to how we feel.

Facts and figures

Teaching is considered one of the world's most stressful professions (McCallum et al., 2017), with higher levels of occupational stress than among the general population in Australia, the United Kingdom and the United States (Bailey, 2013; Education, 2014; Milburn, 2011).

Causes of teacher stress are complex and varied, and the high rate of incidence causes serious concern (Howard and Johnson, 2004). High rates of teacher attrition are being seen around the globe, but 69 million teachers are needed worldwide to reach universal basic education by 2030 (UNESCO,

2022). Retention can only be achieved by addressing the lack of congruence between expectations and reality (Curry and O'Brien, 2012).

Fifty per cent of respondents to a teacher wellbeing report by TES Australia (2022) said their workload was unmanageable. Forty-four per cent said they didn't have a voice on how things were done. Thirty-seven per cent wouldn't recommend their school to friends as a place to work. Only twenty-five per cent said their leadership team made good decisions and 49 per cent felt valued at work.

Research by the Australian College of Educators showed that 28 per cent of teachers work more than a five-day week, with 40 per cent dedicating more than 10 weekly hours to administration. This results in a poor work-life balance for 75 per cent of teachers. Only 33 per cent of those surveyed were operating at their best, with 75 per cent saying they felt stressed at work most of the time (NEiTA, 2021).

A similar 2022 report by Education Support UK found that 72 per cent of educators and 84 per cent of school leaders describe themselves as stressed. Seventy-four per cent of educators found it difficult to switch off and 57 per cent had considered leaving the profession over the previous two years. Forty-two per cent used food as a coping mechanism, while 32 per cent used alcohol. In a study by researcher Jonathan Glazzard of Leeds Beckett University's Carnegie School of Education, 77 per cent of respondents claimed that poor teacher mental health was detrimentally affecting student progress. Ninety-four per cent said that their energy levels in the classroom dropped during periods of poor mental health, and that their teaching was less creative during these times (2018).

Further reports have indicated that 20 per cent of teachers feel stressed about their job most of the time, compared to 13 per cent of similar professionals (Worth and Van Den Brande, 2019). In the USA, only 39 per cent of teachers are satisfied with their profession (Met Life Insurance company, 2013). Fifty-five per cent of American teachers report low morale (National Union of Teachers, 2013) and 15 per cent leave the teaching profession every year (Seidel, 2014).

In China, teachers have a lower health status than the general population with a higher prevalence of anxiety, hypertension, headaches, psychosomatic disorders and cardiovascular diseases (Yang et al., 2009). A lower quality of life and a shorter life expectancy for teachers have also been reported, and this has been attributed to their higher occupational stress (Yang et al., 2009)

Teacher stress can also impact the quality of social and emotional interactions in the classroom. Positive classroom climates are cultivated when we are warm and caring, sensitive to student needs, able to listen deeply and refrain from using sarcasm and harsh disciplinary practices (Reyes et al., 2012). Poor school connectedness results in poorer academic and mental health outcomes for students (Bond et al., 2007). Teachers who are engaged, committed and enjoy their work provide greater support to their students by encouraging intrinsic motivation, self-regulated learning and higher achievement (Watt and Richardson, 2013).

How we give feedback can also impact student outcomes. Emotionally exhausted teachers may use reactive and punitive responses that contribute to negative classroom climates and student-teacher relationships (Osher et al., 2007; Yoon, 2002). When we have the skills to manage daily stressors, we create a safe, respectful and supportive environment that facilitates motivation and learning.

What poor wellbeing looks like

Teachers are givers. We give our time and energy willingly to ensure that others have what they need. But when the demands become overwhelming, it's time to acknowledge that your resource ratio is out of balance.

Teachers who have little left to give are at risk of entering what I call the seven dire states of teacher wellbeing:

1. Disconnection

We shut down when we feel overwhelmed by information and constant change. Becoming less open to growth and new ways of thinking can stifle improvement (Tschannen-Moran and Gareis, 2015).

2. Disengagement

When we are emotionally exhausted, we disengage from what is most important to us. When our needs and values do not align with our environment, heightened emotions can lead to reactive and punitive responses that contribute to poor student-teacher relationships (Osheret, 2007).

3. Detachment

We care deeply about our work. When our expectations are not met, we take it personally. We detach emotionally, reduce work goals and alienate ourselves (Richardson, Watt, and Devos, 2013).

4. Dutifulness

We often act out of obligation, which can breed resentment. If we cannot reflect our idealised image of perfection, then threats to our self-worth may further reduce efficacy (Parker and Martin, 2009).

5. Disappointment

We look to others to feel supported or appreciated for our hard work, yet are left disappointed when people don't meet these needs. This disappointment can lead to irritability expressed at home and in the classroom (Howard and Johnson, 2004).

6. Devaluation

We often feel that what we give is not adequately rewarded. Teachers feel undervalued by high-stakes accountability policies with a lack of reward or recognition (von der Embse et al., 2016).

7. Demoralisation

We feel helpless because we are not consulted in decision-making. A repeated lack of autonomy can cause people to stop trying in the face of adversity. This is known as learned helplessness (Seligman, 2001).

➔

OVER TO YOU

Do you experience any of the dire states of wellbeing? How frequently do you deal with these feelings?

Take this short inventory of your wellbeing at work. Is it in the green zone, the yellow zone or the red zone?

Statement	1 = very poor, 5 = very good				
My sense of belonging to my school community	1	2	3	4	5
My ability to complete my work in the time given	1	2	3	4	5
Access to information about how to manage stress or promote self-care	1	2	3	4	5
The sense of meaning and impact I feel I create	1	2	3	4	5
My impression of being valued and appreciated	1	2	3	4	5

Scoring:

Green zone: 21–30

Yellow zone: 11–20

Red zone: 0–10

Understanding the causes of teacher stress

There are many factors that contribute to the stress of teachers. A national snapshot of the US found the top three sources to be a feeling of overcommitment, the demands of needy students and little or no time to relax (Richards, 2012). These stressors led to feelings of exhaustion, deflation, overwhelm and negativity.

We can begin to address these causes of stress by analysing the connections between them. Let's start by breaking them down into three layers identified by the OECD Teacher Wellbeing framework (2020).

1. Personal factors

These are our individual decisions about nutrition, sleep and exercise, of which I call the '**ME**' factors. They also include personal expectations and our ability to set boundaries. Personal factors are driven by how we think, how we talk to ourselves and how we feel about ourselves. Do you prioritise self-care? Do you take time out to rest, recover and restore your wellbeing? The good news is that these personal factors are the area over which we have the most control.

2. Situational factors

These relate to relationships and how we connect, of which I call the '**WE**' factors. Relationships with students, parents and colleagues can be either highly demanding or a wonderful resource. We ourselves can be a person of support or a roadblock. While we cannot control others, we can certainly influence our relationships.

3. Professional factors

This refers to the administration, compliance, marking, record-keeping, report-writing and other 'stuff' we all have do, of which I call the '**US**' factors. Although it can be strongly influenced by the system in which we teach, this aspect of the job is pretty universal. The systems and processes that make up our workload are often the most demanding with the least opportunity for control.

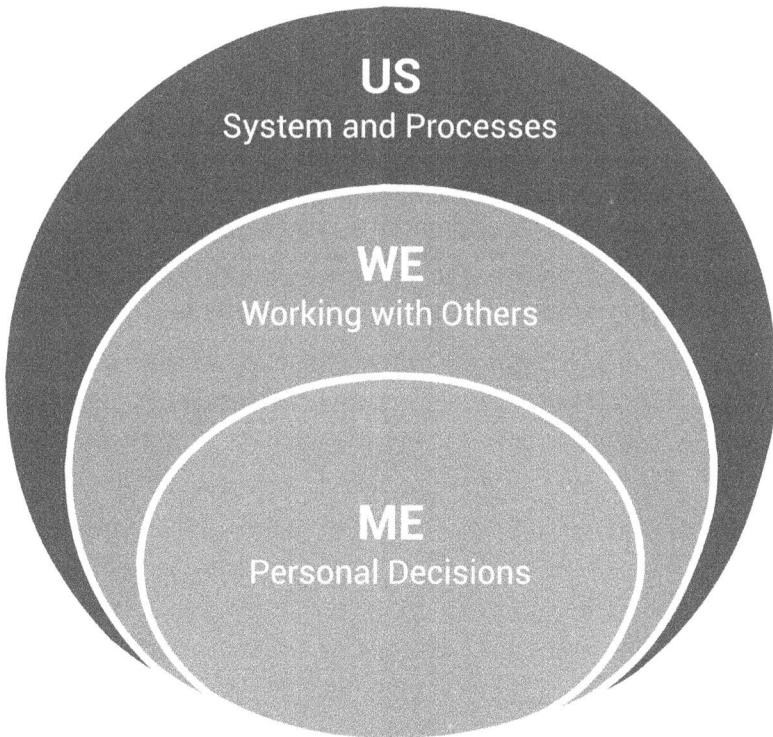

The interplay of wellbeing factors

Our personal wellbeing influences how we connect with others, which in turn impacts our organisation. Demands within the system may impact relationships and therefore negatively affect us as individuals. It is this reciprocal connectedness between each layer that makes the question of teacher wellbeing so complex.

Teachers are not solely responsible for their wellbeing at work. There are larger socio-cultural forces at play. It doesn't matter how much someone exercises or meditates if systemic demands are unreasonable. The environment in which we operate is not solely responsible either. Organisations and their leaders cannot 'do wellbeing' for others. No-one can make you physically fit. No-one can restore your energy if you are feeling lethargic after choosing to eat six muffins at morning tea. We do need to take some responsibility for our own wellbeing.

School leaders who want to support the wellbeing of their staff often share mindfulness apps and offer coffee carts, which aligns with the *personal factors* category. Yet teachers want resources and support with administration, which sits in the *professional factors* category. Here lies the disconnect and frustration. Schools are trying to solve professional issues with personal solutions. By incorporating wellbeing strategies that focus solely on personal factors, we are saying that wellbeing is solely an individual responsibility. The reality is wellbeing at work is a shared responsibility between our personal choices, the relationships we encounter and systemic demands.

While much of what we are asked to do within the education systems is non-negotiable, there are some things that we can control. One way to unpack our levels of autonomy is to explore the balance between demands and resources.

Demands and resources

The WHO defines work-related stress as an imbalance between the demands of a job and the resources available to meet these demands. Every day we are on a seesaw between all the things we have to do and the resources we have to do these demands.

Since its introduction, the Job Demands-Resource (JD-R) theory (Bakker and Demerouti, 2001; 2007) has become an increasingly popular framework for understanding employee wellbeing. The JD-R model takes a dual-process approach to harm reduction and wellness promotion based on employee workplace experience.

Job demands can be physical, social/organisational and psychological. They require sustained investment of energy at a cost to the individual (Skaalvik and Skaalvik, 2018). Job resources are the physical, social, organisational or psychological aspects of the job that stimulate personal growth, learning and development to support the achievement of goals (Bakker et al., 2007). It goes without saying that if demands are high and resources are low, people will feel depleted and productivity declines. If we strengthen resources, people are better placed to navigate daily demands (Bakker, Demerouti and Euwema, 2005).

JD-R theory aims to not only prevent burnout by reducing demands, but also to boost engagement through the development of resources.

Let's discover the demands that impact our wellbeing through the lens of the three layers we just discussed.

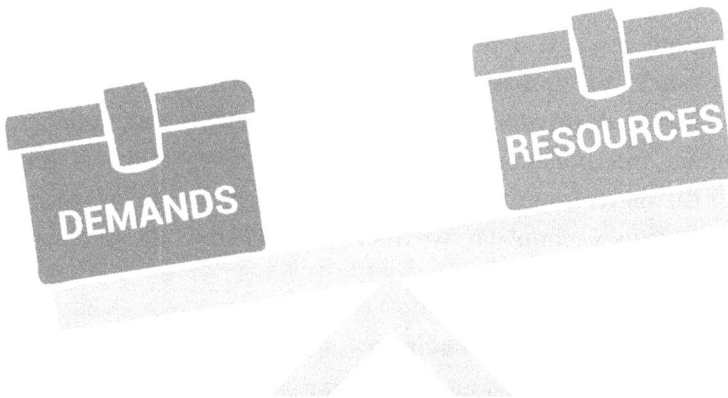

Personal factors: The *ME* stuff

Poor self-care

It's been established that teachers are more likely to suffer sleep disorders, forgetfulness and irritability than non-teachers (Scheuch, Haufe and Seibt, 2015). This is because we tend to prioritise the wellbeing of others. Poor exercise and nutrition routines compound poor sleep habits, which depletes resources. I remember weekly staff morning teas that resembled the food you would get at a child's birthday party. These foods high in fat and sugar impaired cognitive functioning for the rest of the day. The decisions we make about what we eat during a workday directly affect our psychological and energy resources.

Poor emotional regulation

Teaching is an emotional vocation. On a single day you can experience a spectrum of feelings ranging from anger to joy and despair to pride. We don't just juggle our own emotions; from students throwing temper tantrums to parents with separation anxiety, others expect us to manage their moods, expectations and behaviours. We must remain professional by concealing high emotions that overstimulate our nervous system and instigate a stress response in our body. If we don't calm this stress response regularly, we risk energy depletion and burnout at the end of each semester. This means getting the flu in the holidays because you are exhausted.

Emotional regulation skills are a key intrapersonal resource that protects against the cumulative effects of stress (Jennings and Greenberg, 2009). Poor emotional regulation is closely associated with mental health problems including anxiety and depression, while ineffective emotional regulation strategies such as suppressing or faking emotions have been associated with emotional exhaustion among teachers (Chang, 2013). Emotional regulation can impact interpersonal relationships, including how we respond to students no matter how much we care about them. If we cannot regulate our emotions during a school day, we risk negatively affecting students and their learning.

Unrealistic expectations of self

We develop strong ideas about what it means to be an educator and wear this badge with pride. After all, we chose our profession because we wanted to make a difference in the world. The problem is that we often don't know how to rest. If we do, we are riddled with guilt. After investing our energy in meeting the needs of others, we can also take critical feedback personally. Every day that we walk through the school gate we risk emotional exposure and uncertainty. The constant drive to improve, grow and achieve leads to a reoccurring question: 'Am I doing enough?'

Constant external accountability drives this question further. 'Am I enough?' we ask ourselves. The pressure to measure and show evidence of improvement can lead to comparison and competition. We may not say it out loud, but we secretly want people to notice that we are doing a good job. We desperately seek validation for our work. We don't want fanfare or trophies, but we would like to be acknowledged. One of the best things we can do for ourselves is to acknowledge the incredible work we do instead of striving for unrealistic ideas of perfection.

Decision fatigue

Did you know that we make about 1500 decisions a day? Schools are dynamic places that require extensive decision-making. Decision fatigue arises from the need to constantly make micro and macro decisions on any number of topics: getting through a crowded curriculum, remembering your playground duty, knowing what to say to calm student tempers, giving specific and immediate feedback, modifying a lesson on the spot. When you go home to your family and they ask you what you want for dinner, you'll reply that you don't know and don't care! Being able to rest our minds is crucial.

Low self-efficacy

Self-efficacy is the internal belief that we have the ability to succeed. A study of Norwegian teachers linked positive self-efficacy to job satisfaction and negative to burnout (Skaalvik and Skaalvik, 2014). If we cannot reflect our perceived image of success, threats to our self-worth may reduce self-efficacy (Parker and Martin, 2009). Efficacy beliefs drive our focus, which influences how we respond to student behaviour and learning. A belief that we have little influence over our students' potential to achieve tends to become a self-fulfilling prophecy (Donohoo, 2016). When we become stressed, we add pressure by judging ourselves against what a 'good' teacher would do. If we believe in our ability to succeed, we feel more satisfied and are energised to keep going.

A teacher's sense of efficacy influences their willingness to experiment with new methods. When our self-efficacy is low, we shut down new ideas that could better meet the needs of students (Tschannen-Moran and Gareis, 2015). Teachers with high self-efficacy successfully meet the rising demands of students, are willing to stretch themselves and grow, persist in the face of challenges, foster learning autonomy and convey high expectations (Donohoo, 2016). Self-efficacy beliefs influence our willingness to engage in professional development activities and to seek opportunities for growth and learning (Skaalvik and Skaalvik, 2013). Teachers with higher self-efficacy are more likely to actively pursue professional development, engage in reflective practices and continuously improve their teaching skills.

Positive feedback is a powerful way to influence efficacy beliefs. We grow in confidence when we're celebrated for good work or reflect on our mastery. Confident teachers are open to new experiences and share what they have learned with colleagues. Given the nature of schools as a performance arena, teachers can feel vulnerable having to measure and rate their performance (Parker and Martin, 2009). The external pressures of accountability can directly impact self-efficacy and self-worth, known determinants of resilience and wellbeing (Parker and Martin, 2009). What we do know is that when teachers are given opportunities to observe and positively reflect on their actions, they build their confidence and efficacy (Hattie and Yates, 2014).

Life outside school

Let's not forget that we have lives beyond the schoolyard. Sometimes the demands of life at home outweigh those at school, resulting in tug-o-war. Whether it is supporting our own children, managing elderly parents, attending to house issues or engaging in hobbies, the juggle can be draining. This only adds to our feeling that there is not enough time to get everything done. When things start to fall through the cracks, we rely on our internal resources to find solutions. We feel guilty for not doing at enough at work and for not being our best selves at home.

Situational factors: The *WE* stuff

Interaction

Human interaction is a necessary ingredient in education. Colleagues, students, parents: we're all in the same storm, just in different boats. The quality of our interactions determines whether we feel supported, valued and recognised, or undermined and alone. Being constantly available to others means that we are engaging in an estimated 1000 interpersonal contacts a day (Holmes, 2005). These interactions can affect our beliefs about our ability, which can deplete emotional resources.

Colleagues

Elements of our individuality can lead to misunderstandings, miscommunications and ill-informed judgements that result in personality clashes between colleagues. Feeling that we have not been acknowledged can lead to feelings of rejection and disconnection. A lack of social support can contribute to high levels of stress and burnout (Kaihoi, Bottiani and Bradshaw, 2022). Acknowledging our differences and creating inclusive practices improves connection and belonging.

Providing social support is a significant task for those in senior roles. It is the variable with the greatest likelihood to increase job satisfaction and reduce the risk of burnout (Bakker and Bal, 2010). School leadership has a significant influence on teacher motivation and engagement (Bird et al., 2009). A lack of support from leadership is associated with anxiety disorders, mood disorders, depressive disorders and neurotic disorders (Day and Gu, 2013). Workplace demands become heavy if leaders do not consciously promote caring and supportive relationships to ensure that staff feel valued, respected and supported (Bryk and Schneider, 2003). This lack of feeling cared for influences perceptions of engagement and ultimately the climate of the organisation (Ahghar, 2008).

From my experience, while schools do have that five per cent of people who are described as 'difficult to work with', most staff proudly share that they like each other, are mutually supportive and enjoy bantering with colleagues. These relationships form the basis of what is called social capital, one of the key pillars in shaping teachers' professionalism (Hargreaves and Fullan, 2012). In fact, teachers who have low knowledge and skills but high social capital have better outcomes than high-ability teachers with low social capital (Leana, 2011). High social capital builds trust and is crucial for occupational wellbeing. When social capital is low, we tend to operate in silos and miss out on the benefits of collaboration.

Illustrating the phenomenon known as contagion effect, studies have shown that individual burnout levels are uniquely related to the average burnout level in a school. This reinforces the understanding that colleagues can have either a positive or negative impact on the stress levels of others (Kim, Youngs and Frank, 2017).

Students

We love our students, but we'd be foolish to think it's easy to meet their complex needs and our own at the same time. Engaging lethargic and unmotivated learners can be challenging. Sometimes communicating with young people can feel like talking to a brick wall. It's hard not to get frustrated when students don't hand in assignments after we've offered hours of extra support. We judge our impact by the grades our students earn, and are disheartened when they're bored after we've spent hours preparing a lesson.

A significant body of research suggests that student behaviour can have a substantial impact on teacher stress and wellbeing. It's clear that teacher perception of student misbehaviour is significantly related to emotional exhaustion and job dissatisfaction (Hakanen, Bakker and Schaufeli, 2006). Teacher stress is directly affected by student behaviour problems, with teachers reporting higher levels of stress when dealing with students who exhibited disruptive or aggressive behaviours (McCarthy and Lambert, 2006). Teachers who experience more student discipline problems report lower levels of job satisfaction and higher levels of burnout (Stoeber and Rennert, 2008).

Parents

Parents are generally highly invested in meeting their child's needs, and can take aim at teachers when these needs are not met. When parents are critical, hostile or uncooperative, teachers experience increased stress, burnout and job dissatisfaction (Boynton-Jarrett et al., 2013; Chen and Li, 2017). This may be particularly true for those who work in schools with a relatively low socioeconomic status, where parents may be more likely to express negative attitudes towards teachers. On the other hand, some parents may be overly involved in their child's education and become overly critical or demanding of teachers.

One challenge is finding the time to meet parents. While parent-teacher communication is key to supporting students, the quality of this relationship can also affect the wellbeing of teachers. Effective communication between parents and teachers can help build trust, promote collaboration and prevent misunderstandings (Hong and Ho, 2005). When communication is poor or conflictual, teachers may experience increased stress and job dissatisfaction (Boynton-Jarrett et al., 2013).

Supportive, respectful and engaged parents can help create a positive school culture that benefits teachers as well as students. Critical or uncooperative parents may ultimately harm the quality of education that their children receive.

Professional factors: The *US* stuff

Workload

As school systems become more bureaucratic, workloads become more burdensome for everyone in the organisation. High workloads are a significant predictor of teacher stress and burnout (Kim and Lee, 2011). Greater service-delivery expectations are met with fewer resources (McCallum et al., 2017), and little planning time is offered. This results in teachers having to blur boundaries between home and work (McCallum et al., 2017). Combining with an ever-increasing emphasis on accountability measures is the exclusion of teachers from policy-making procedures (McCallum et al., 2017). The results of this imbalance are frequent turnover, low performance, absenteeism and efficiency costs (Albulescu and Tuşer, 2018).

Closely related to this are the multiple jobs that teachers are expected perform outside of their teaching roles: extra-curricular activities, school management, communicating and cooperating with parents, providing counsel to students. These tasks place extra pressure on us and negatively impact our sense of professional wellbeing (Valli and Buese, 2007). With schools increasingly becoming places of social and political awareness, the unrealistic expectations placed on teachers to meet related demands rise (McCallum and Price, 2010).

Lack of control

Educational bureaucracy demands that teachers follow instructions, yet we're also expected to be creative and innovative to meet the diverse needs of our students. Too often, our autonomy is undermined by administrative control, inflexible curricula and a lack of resources.

Participating in decision-making at a school level is significantly related to job satisfaction across all countries (OECD, 2014). Work autonomy is considered a basic psychological need. Teachers should be given sufficient autonomy to make decisions about how to best do their job (Skaalvik and Skaalvik, 2009). Teachers without autonomy feel micromanaged, which can lead to learned helplessness. As demands grow, feelings of being at the mercy of others can turn to cynicism and resentment.

Lack of rewards

Teachers benefit from being recognised and valued yet systemic performance reviews all too often leave us feeling judged and undervalued. Research shows that positive appraisal and feedback help teachers cope with job demands (Bakker et al., 2007). The teacher appraisal and feedback system can have a positive effect on job satisfaction if perceived to be conducted in a fair and respectful manner (Vanhoof et al., 2014).

It's essential to ensure that high-quality teaching is occurring in our classrooms, but performance evaluations can be a source of stress, burnout and ill-health. High-stakes accountability policies positively correlate with teacher stress and lower job satisfaction (von der Embse et al., 2016). One reason for this could be the lack of reward or recognition accompanying these reviews.

⊙

OVER TO YOU

- Take a moment to reflect on the demands that negatively impact your wellbeing at work. Categorise these into the areas of **ME**, **WE**, **US** factors.
- Now that you have categorised the types of demands placed upon you, which area do you have most control?

What it means to thrive

Organisational psychology expert Gretchen Spreitzer describes thriving as a psychological experience comprised of both vitality and learning—the antithesis of burnout (Porath, Gibson and Spreitzer, 2022). To thrive is not merely to survive, but to willingly grow.

People who thrive at work tend to be healthier, more engaged and better-performing, with a buffer against stress and negativity. They report less burnout (Spreitzer, 2005), have better health with fewer sick days (Spreitzer, 2012), experience better job satisfaction and are more committed to their organisation (Porath, 2012). Thriving people are more open to learning and taking on new initiatives at work (Quinn, 2016) and collaborate more with others by sharing resources (Spreitzer, 2016).

People who thrive at work report less burnout because they are able to generate resources rather than deplete them. Thriving makes us more likely to experiment with new ideas and seek out new ways of working. When people feel energised, they engage in proactive behaviours helping them better manage daily challenges. Thriving people generally feel they are co-creating their environment, which builds further resources such as high-quality connections (Spreitzer, G., et al., 2004).

➡️

OVER TO YOU

- What might it look like if someone was thriving at work?
- Can you recall a time when you were thriving at work or in your personal life? What were you doing? How did you feel?

Learning to thrive as an adult

Adults have knowledge, diverse interests and lived experience. We generally don't like being told what to do and prefer to make our own decisions. These decisions will have an impact on your ability to thrive.

The challenge for teachers is that professional learning typically focuses on building knowledge related to the curriculum or subject matter. How often are we enrolling in courses to develop our sense of identity? In what ways do we learn resilience skills? How often are we challenging our thinking patterns to see if we are helping or harming ourselves? Because schools are time-poor, little effort is afforded to teaching us how to thrive. While there is no one-size-fits all model, research can provide us with evidence-based strategies.

One of the biggest mistakes I see is leaders using their student wellbeing frameworks to address staff. Not only does this approach not work, but it disrespects adult learners and their need for autonomy.

The Australian Institute for Teaching and School Leadership (AITSL) has identified several key themes to ensure that professional learning is effective for teachers. These include allowing teachers to set their own goals, providing ongoing learner-led exploration of issues and having a flexible approach to continuous, collaborative learning.

To gain the skills to thrive, we need to follow these five adult learning principles:

1. Be a self-directed learner

You are encouraged to actively engage by identifying your learning needs, setting goals and determining the best strategies to acquire new behaviours. This autonomy and involvement in your own learning enhances motivation and ownership, leading to more effective and sustainable changes.

2. Consider relevance and practicality

Adult learners seek practical knowledge and skills immediately applicable to their professional context. Aligning new behaviours with the specific challenges you face increases the relevance and value of your learning experience. Incorporating real-world examples, case studies and opportunities for reflection and application will help you see the direct impact of adopting new behaviours and reinforce your commitment to change.

3. Value prior experience and expertise

By acknowledging and leveraging your background, you can tap into your existing knowledge and skills to make connections with the new behaviours you want to learn. This approach not only enhances understanding but also boosts confidence and self-efficacy, as you recognise that you already possess valuable resources to support your growth.

4. Explore collaborative and experiential learning

Opportunities to engage in discussions, share ideas and collaborate with peers foster a sense of community and mutual support. Through interactive activities, role-playing and simulations, you can actively experience and apply new behaviours in a safe and supportive environment. This experiential approach enhances comprehension, skill development and the transfer of learning to real-world teaching situations.

5. Continuous professional development

A culture of continuous improvement is fostered by reflecting on activities, seeking ongoing learning opportunities and embracing a growth mindset. By emphasising the value of lifelong learning, I want to empower you to remain adaptable, open to change and committed to strengthening your wellbeing throughout your career.

A message for leaders

Leaders have a responsibility to be the gardeners of their teams. Your job is to know the plants in your garden, plan according to the seasons and adjust to conditions as they occur. This means identifying the fragile seedlings such as early-career teachers who need extra support and care. You must also take time to appreciate the rose bushes, beautiful and resilient despite their prickles. These people are generally longtime educators who have wisdom to offer, though scars from harsh conditions have left them a little weathered. You must be vigilant in stamping out the weeds of gossip, hearsay and negativity.

When I work with school leaders, I give them this advice: stop running boring meetings! Teachers are professional engagement officers. We are skilled in getting the attention of easily distracted students. When it comes to educating colleagues, however, we revert to 'show and tell'. Don't fall for the myth that we don't have time to do things differently. I push back on this argument and urge leaders to be more creative. Leaders are experts in education, with both wisdom and experience. Take the time to build on these qualities by improving your understanding of adult learning principles. Be brave in trying new ways of sharing knowledge, and build your leadership capacity so we can move beyond compliance to care.

A highly effective field that can support you in creating these conditions is coaching psychology. Coaching skills can help leaders better understand how people think, as well as how to best develop their potential. When armed with strategies to empower others, you will move from the position of 'sage on the stage' to that of 'guide on the side'. By shifting how we communicate, we can enable conditions in which people can grow and thrive at work.

Conclusion

Supporting wellbeing at work is complex. Your personal wellbeing (**ME**) may influence how you connect in relationships with others (**WE**), which then could impact the organisation (**US**). Alternatively, the demands within the system (**US**) may impact relationships at work (**WE**) and therefore have a negative impact on individual teachers (**ME**). It is this reciprocal interconnectedness between each layer that needs to be understood if we are to plan meaning ways for us to thrive.

The good and bad of stress

The stress response

Stress is an inevitable part of life, a natural response to any situation that challenges our ability to cope with the demands placed upon us. Associated with negative emotions such as anxiety and frustration, it can also be a positive force that motivates us to take action and achieve our goals.

Prolonged or chronic stress can have a detrimental impact on our physical and mental health. Stress prompts the release of hormones such as cortisol and adrenaline, which trigger a cascade of physiological changes to help us respond to the perceived threat. These changes can include an increased heart rate, elevated blood pressure and the release of glucose into the bloodstream to provide energy. The body also redirects blood flow to the muscles and away from non-essential organs such as the digestive system.

While these responses are designed to be helpful in the short term, chronic exposure to stress can result in a range of negative health outcomes including high blood pressure, an increased risk of heart disease and a weakened immune system. Prolonged stress can also have a significant impact on our mental health, contributing to conditions such as depression, anxiety and burnout.

Given the negative impact of chronic stress, it is important to develop effective strategies for preventing it. This chapter will explore the causes and consequences of stress, as well as the various approaches that we can take to combat it.

Real-life example

Dina was a dedicated teacher who cared deeply about her students and their success. She spent long hours preparing lessons, grading assignments and providing extra help to struggling learners. Despite her best efforts, Dina felt stressed and overwhelmed by everything she had to do. She found herself struggling to sleep and becoming irritable and short-tempered with her students, colleagues and family. She began to question if it was all worth it.

During a series of workshops, I spoke with Dina about how she was feeling. We explored the importance of taking your MEDS (meditation, exercise, diet and sleep) through a self-care action plan. With an accountability buddy, Dina agreed to do a walk three times a week before school. She also planned a healthy before-bed ritual complete with mindfulness. Not before long, she found she was better able to manage stressful situations.

Over time, she began to feel more balanced and energised. Although the demands on her were still high, she was able to connect with her students and colleagues in a more positive way. She learned that self-care wasn't about being selfish, but about being responsible. If she was going to be her best self for her students, she needed to prioritise ways to manage the daily stressors she experienced.

Stress is the new normal

With so much change and uncertainty in today's world, stress has become an everyday experience. While technology has helped us make life easier in many ways, it has also left little opportunity for rest. We live on high alert in a 24-hour society that expects us to be available to all who need us. We are more connected than ever, yet can feel disconnected and lonely. We strive to be more, do more and give more with little attention to our physiological needs. We push our bodies to run as high-functioning machines to the point where exhaustion is the norm.

We are addicted to being busy, even using the word to symbolise how hard we are working. In a case study published in *Educational Psychology*, a teacher described her compulsion to be busy and how it negatively impacted her life and work. She stated: 'I feel like I'm always behind, and I'm constantly trying to catch up. I'm addicted to being busy, and I don't know how to stop' (Marshall, 2015, p. 219).

We validate each other for remaining busy and judge each other for not being busy enough. I remember seeing a colleague sitting in the staffroom reading the newspaper at lunchtime one day. I was shocked. How could this person possibly have time to sit and read the paper? Surely he could be completing unfinished tasks or preparing his lessons to be more engaging? In my world, if you weren't rushing from one thing to the next then you simply weren't doing your job well enough. Living under high stress was not only the norm; it was expected.

How stress occurs in the body

Humans are designed to experience stress and react to it. When we're under pressure, chemicals and hormones surge throughout the body. The physical, emotional and intellectual reaction to this is known as our stress response. This response is the result of the autonomic nervous system (ANS) becoming activated. The ANS is the internal system that controls involuntary reflexive actions such as heart rate, respiratory rate, pupil dilation and digestion.

The physical changes brought about by the stress response enabled our ancestors to perform at their peak and maximise their chances of survival. The problem is that the ANS cannot distinguish between a physical threat and a psychological one. In modern society, the stress response is triggered by a wide range of situations that are not necessarily life-threatening. For example, we experience stress in response to work deadlines, financial pressures, relationship problems and even traffic jams.

This constant activation of the stress response can have negative effects on our physical and mental health, particularly when it is chronic or prolonged. High levels of stress hormones such as cortisol can damage the body's tissues and organs, increase the risk of heart disease and stroke, weaken the immune system and reduce the efficiency of our digestion.

Learning to manage stress through techniques such as relaxation, exercise and mindfulness can help mitigate its negative effects and promote wellbeing.

Signs of stress

It is important to recognise the following signs of stress in ourselves and others in order to take appropriate measures.

Physical symptoms

Stress can affect the body through muscle tension, headaches, fatigue, insomnia, digestive problems and a weakened immune system. People who are under stress may also experience changes in appetite, weight or sex drive, as well as increases in heart rate, blood pressure and sweating.

Emotional symptoms

Stress can cause anxiety, irritability, frustration, overwhelm and depression. You may have difficulty concentrating, making decisions or remembering things. In some cases, it can trigger or exacerbate mental health conditions such as anxiety disorders, depression and post-traumatic stress disorder (PTSD).

Behavioural symptoms

Stress can lead to behavioural changes such as withdrawal, avoidance, aggression and substance abuse. People who are stressed may also have trouble sleeping, socialising and enjoying hobbies. They may neglect aspects of self-care such as exercise, hygiene and medication. They may engage in unhealthy coping mechanisms such as overeating, smoking and drinking, which can further worsen their physical and mental health.

Cognitive symptoms

Stress can affect the way you think and perceive the world around you. You may feel self-doubt, guilt or pessimism, and engage in catastrophic thinking by imagining worst-case scenarios. Your perceptions of reality can also distort. Symptoms include hypervigilance (the perception of threats where there are none) and emotional numbing (disconnection from feelings and surroundings).

Interpersonal symptoms

Stress impacts relationships and communication. You may become more conflictual, defensive or critical, perhaps withdrawing from social interaction altogether. You might also have difficulty expressing your emotions or needs, and may misinterpret the intentions of others. This can further worsen your symptoms and contribute to a negative cycle of interpersonal stress.

When stress becomes burnout

Stress and burnout are related but distinct concepts. While stress is a normal response to challenging situations, burnout is a state of chronic stress characterised by emotional exhaustion, cynicism and reduced personal efficacy. It wasn't until 2018 that burnout was included by the WHO in its *International Classification of Diseases*. The WHO describes burnout as 'a syndrome conceptualised as resulting from chronic workplace stress that has not been successfully managed' (2018). Burnout can occur when the demands of a job or other responsibilities exceed a person's ability to cope with them over an extended period of time.

There are several differences between stress and burnout. Stress is a short-term response to a specific situation or event, while burnout is a long-term state of chronic stress that persists over a period of weeks, months or even years. Stress can cause a wide range of physical, emotional and cognitive symptoms such as headaches, irritability and difficulty concentrating. Burnout is often associated with chronic work-related stress, but can also be caused by factors such as caregiving responsibilities, financial stress or chronic illness. People with burnout may feel emotionally drained, detached from their work or other responsibilities, and unable to make a meaningful impact.

Stress can have negative consequences on physical and mental health, but these symptoms usually go away once the stressor is removed or the situation improves. Burnout, on the other hand, can have long-term consequences on health, job performance and personal relationships. It can lead to depression, anxiety and other mental problems, and contribute to job dissatisfaction, absenteeism and turnover.

Burnout doesn't just happen; it results from slowly eroding coping skills. When work demands don't match people's knowledge, skills or ability to cope then the risk of burnout heightens. Burnout is stress to a saturation point, and it can creep up on you without notice.

Burnout and teachers

The pressures of managing large classes, dealing with complex curricula and navigating administrative pressures make teachers susceptible to burnout. A study found that teachers reported significantly higher levels of burnout than people in other professions, with emotional exhaustion being the most commonly reported symptom (Elfering, Gerhardt, Grebner and Muller, 2015). High workloads, low social support and low autonomy were the strongest predictors of burnout among teachers.

Another study found that teachers who perceived their work as emotionally demanding were more likely to experience burnout (de Jonge, van den Berg and Tummers, 2018). This study also found that burnout was likely to be provoked by deficits in job resources such as social support, work autonomy and feedback from colleagues.

The COVID-19 pandemic and subsequent move to virtual lessons placed significant additional stress on teachers and increased their risk of burnout. A study by Li and colleagues (2021) found that teachers reported increased levels of burnout and reduced job satisfaction, with those teaching online reporting the highest levels of burnout.

Stressors and stress responses

It is important to understand the distinction between stressors and the stress response. Stressors can be tangible (a challenging workload) or intangible (perceived threats or worries). They can be either internal or external. While external stressors are often out of our control, internal stressors are more likely to be within our control.

External stressors include student behaviour, parental expectations and administrative demands. Some of the biggest internal stressors faced by teachers are our expectations of ourselves. We often set high expectations for student achievement, classroom management and instructional delivery. Striving for perfection and feeling the need to meet unrealistic standards can create internal stress and self-imposed pressure.

For me, common stressors were 'that' parent who was always in your face, 'that' child who just wouldn't do what they were told and 'that' colleague who seemed to roadblock every decision you wanted to make. I dreaded talking to these people. My face would go red, my heart rate would speed up and adrenaline would run through my veins. I would have to use all my psychological strength to bottle up my emotions, nod politely and walk away.

➡

OVER TO YOU

- Recall a time you felt stressed at work. How did your body react? How would you describe your heart rate, breathing, skin temperature?
- How did you respond to the situation? Did your actions aggravate or calm your nervous system?

Closing the stress cycle

Just as we can adjust the intensity of light in a room by turning down the dimmer switch, we can also regulate our stress response. The parasympathetic nervous system, also known as the 'rest and digest' system, acts as a counterbalance to the sympathetic nervous system by helping the body return to a state of calm and relaxation. This response slows down the heart rate, reduces blood pressure and allows the body to recover.

I successfully managed the stressor of conversations with 'difficult' students, parents and colleagues by not engaging in arguments, but my body remained in a heightened stress response. I needed to find a way to calm my nervous system down and let my body know I was safe. Strategies to activate my parasympathetic nervous system would allow my body to return to a neurological and physiological balance. While I had little control over the stressor (the person), I had to take responsibility for managing my stress response.

High levels of stress can cause the nervous system to become stuck in the sympathetic response, even when we don't need it to be. To alleviate constant symptoms of stress, we must restore our equilibrium.

In their 2019 book *Burnout*, Emily and Amelia Nagoski refer to this process as 'closing the stress cycle'. This means allowing the body to fully recover from the physiological and psychological effects of stress, rather than allowing these effects to linger or accumulate over time. Closing the stress cycle can increase resilience to future stressors (Maslach and Leiter, 2016).

By allowing the body to fully recover from the effects of stress, we can better cope with future stressors and maintain a sense of balance and wellbeing in our daily lives.

Strategies that can help close the stress cycle include exercise, deep breathing, mindfulness and social support (McGonigal, 2015). Exercise has been shown to be particularly effective in reducing the physiological effects of stress such as increased heart rate and blood pressure, while mindfulness and social support can help reduce psychological effects such as anxiety and depression. We'll talk more about these strategies later in the book.

OVER TO YOU

- Think about how you experience stress at work. When your nervous system becomes heightened and you can feel your heart rate and blood pressure rising, how do you return yourself to a normal state?
- What activities do you find calming or relaxing after a busy day?

Your perception matters

While we may not be able to directly influence most of our external sources of stress, we can certainly influence our internal ones. This is because our physiological response is directly related to our perception of the presence or absence of threat. In other words, stress is a subjective experience that varies from person to person.

Imagine two people faced with giving a presentation at work. One person might perceive this as a challenging but manageable task, while the other might perceive it as overwhelming and stressful. The stress response is not caused by the presentation itself, but rather by the individual's perception of the task.

Factors that can influence our perception include our beliefs, values, past experiences and coping strategies. If we believe that we are not capable of handling a particular situation, we are more likely to experience stress. If we have effective coping strategies and a positive outlook, we may be better equipped to handle stressful situations without feeling overwhelmed.

It is important to recognise that our perception of a stressful event is not always accurate or helpful. Sometimes we may perceive a situation as more stressful than it actually is, or see it as hopeless when there are solutions available. This is referred to in psychology as cognitive distortion or a thinking trap. In these cases, it may be helpful to challenge our perceptions and seek support from others to help us cope with the situation.

Recognising stress as a perception of an event can help us understand its subjective nature and the importance of managing our perceptions and coping strategies to maintain our wellbeing.

Our inner critic

Perfectionistic teachers struggle with delegating tasks and seeking help from others. They may feel that they are the only ones capable of meeting their high standards, leading to an increased workload and added stress. This reluctance to delegate can also limit collaboration and teamwork, which are essential for a supportive teaching environment. These teachers may have unrealistically high expectations of student performance, leading to frustration and disappointment when these expectations are not met. This can strain the teacher-student relationship and create additional stress.

To address perfectionism, we need to set realistic and attainable goals for ourselves. We need to practice self-compassion and encourage awareness of our thinking patterns so we can counteract self-criticism. We need to recognise that mistakes are a natural part of the learning process and treat ourselves with kindness and respect, just as we do our students. We will also benefit from seeking support from others who understand our situation. Sharing experiences and concerns can provide validation, guidance and a sense of camaraderie, reducing the isolation that perfectionistic tendencies can create.

Encouraging personal boundaries between work and your personal life is also vital. School leaders should promote a culture that values effort, growth and collaboration over perfectionism. Encouraging open communication, providing resources for professional development and recognising achievements can contribute to a more supportive and less stressful environment.

OVER TO YOU

- What are some of the phrases that your inner critic says to you?
- Do you or others you know have a tendency towards perfectionism? What does it look like?
- What advice would you give someone to soften their inner critic?

Stress as a positive experience

Stress does have its benefits. It can help us rise to challenges and keep us alert in the face of danger. The technical word for this beneficial stress is 'eustress'.

Psychologist Kelly McGonigal explains in her popular 2016 book *The Upside of Stress* that we can use stress to help us focus on goals instead of threats. She is referring to the stress that drives excitement or motivation. This is the stress you feel when you're about to meet your new class for the first time, or as your students are going on stage in front of parents after 10 weeks of rehearsal.

Instead of draining your energy reserves, this adrenaline can fuel your energy to keep you motivated and focused. Interestingly, McGonigal explains that people's beliefs about stress influence how their bodies respond to it. An eight-year-long study surveying over 30,000 Americans found that highly stressed people who feared that stress was bad for them were 43 per cent more likely to die than any other group. Highly stressed people who didn't think that stress was bad for them had even lower mortality rates than those who led low-stress lives. If you believe stress is bad for you, you are more likely to die from stress-related illnesses such as heart disease and cancer.

What if we looked at our body's stress response as helpful instead of harmful? What if instead of wanting to eliminate stress altogether we were able to harness the energy of stressful events to become happier and more productive?

It's true that long-term stress can lead to a variety of physical and mental health problems. However, it is important to recognise that stress can have positive effects under certain circumstances. Short-term stress, also known as acute stress, is a natural and adaptive response to a challenging or exciting situation such as a job interview, a sports competition or a thrilling activity like bungee jumping. In such situations, the body's fight-or-flight response increases energy, alertness and focus. This can improve performance and lead to a sense of achievement and satisfaction after the event is over. Stress can increase motivation, focus and creativity, leading to personal growth and development. It can be productive in challenges such as learning a skill, taking on a project or overcoming an obstacle.

Stress can have a positive influence on personal growth and resilience. When people face difficult life events such as illness, loss or trauma, they may experience a significant amount of stress. However, they can develop their resilience and adaptability if they are able to cope with this stress in a healthy way by seeking support from others, practicing self-care or finding meaning in the experience. This can lead to a greater sense of self-awareness, personal growth and even post-traumatic growth.

OVER TO YOU

- Describe a time where you felt pumped or excited at work. What was happening? How did it make you feel?
- Is there a way that you can create that feeling more often for yourself?

Conclusion

Stressors and the stress response in our body are different. While stressors are often external and out of our control, our internal stress response is something we can manage both cognitively and physiologically. No-one can remove the stress response or manage it for you. It is something you have to action.

Teachers sometimes look to school leaders to fix our wellbeing. We want magic solutions to reduce our feelings of stress and burnout. We oscillate between blaming others for how we feel and crying in despair. We want more energy; we want to function well and know that we are doing well.

We need to build awareness of our internal and external stressors and learn ways to better manage our stress response. We need to give ourselves opportunities to build our capacity to cope.

Theories in wellbeing

Talking about wellbeing

What does it truly mean to live a life of wellbeing? This fundamental question has perplexed philosophers, psychologists and seekers of happiness for centuries. You may be surprised to know that there is no agreed definition of wellbeing. While the word is thrown around frequently in education, it is a vast concept that is difficult to define. As we navigate the complexities of being human, we need to understand how we can live a truly fulfilling, contented and meaningful life.

In our quest we will explore interconnected theories of subjective, hedonic, eudaimonic and psychological wellbeing, as well as perspectives on flourishing and motivation. By shedding light on the characteristics of these theories, we build a language to better define what wellbeing at work means.

We're very good at having conversations about stress and exhaustion, but not about wellbeing. Let's build our vocabulary to focus on what we want as opposed to what we don't want. With new language we can better articulate our needs and develop our resources to meet them.

Defining wellbeing

An early reference to wellbeing appeared in the *World Health Organization Act 1947*, which defined health as 'a state of complete physical, mental and social wellbeing and not merely the absence of disease or infirmity'. This model has since been expanded to encompass economic, psychological and social dimensions.

In 2004, the WHO defined mental health as 'a state of wellbeing in which the individual realises his or her own abilities, can cope with the normal stresses of life, can work productively and fruitfully, and is able to make a contribution to his or her community'.

From these definitions we see that wellbeing encompasses connection and contribution, highlighting the influence of environmental stressors.

Theories of wellbeing

Educators are well-versed in the importance of drawing on research to guide pedagogy and student wellbeing programs. But when it comes to adult wellbeing, we seem lost. This is because we are experts in educational psychology, but not in organisational psychology that supports human flourishing in the workplace.

We draw on our limited knowledge to plan trivial and tokenistic wellbeing initiatives such as coloured sock day. We copy student wellbeing programs and assume they will work for adults. We mandate weekly morning teas with the intention of building positive connections, yet people use this time to bicker or blame others for decision-making.

If we are serious about nourishing the wellbeing of teachers, we must draw on research to inform our decisions. We must deepen our understanding of wellbeing and what it means to be well at work. We must build psychological, social and emotional resources to better manage demands. We must take the time to learn what it means to thrive.

Hedonic wellbeing

The concept of hedonic wellbeing dates back to Greek philosopher Aristippus in the fourth century B.C. It is based on the notion that increased pleasure and decreased pain lead to happiness. The goal here is to feel good, but it is unrealistic to think we can feel good all the time. Life is full of ups and downs. If we fall into the trap of chasing pleasure, we find ourselves on what is known as the hedonic treadmill.

I tend to liken hedonic wellbeing to a sunset. An incredible sunset is always happening somewhere in the world. It's my favourite part of the day, to sit somewhere and watch the colours change from blue to pink to purple. I love the golden glow that gently caresses treetops and rooftops, the edge of the sun slowly dropping further under the horizon as shadows move and the light fades. It's a moment of short-lived pleasure, peace and stillness. I cannot capture the feeling or store it for later. I cannot hold or control it. It is merely an experience that happens for a moment and then is gone.

➡️

OVER TO YOU

- Which positive emotions are you most likely to experience at work?
- When you feel good at work, what is happening in and around you?

Subjective wellbeing

Subjective wellbeing comes from hedonic concepts, and relates to life satisfaction. The term was coined in 1984 by psychologist Ed Diener, affectionately known as Dr Happiness. Based on Diener's works, the OECD defines subjective wellbeing as 'good mental states, including all of the various evaluations, positive and negative, that people make of their lives and the affective reactions of people to their experiences' (2013).

Central to our quality of life, subjective wellbeing is directly related to emotional wellbeing. People who feel satisfied with their lives and frequently experience good feelings such as joy, contentment and hope are more inclined to be seen as enjoying a high quality of life.

Subjective wellbeing can be measured using a satisfaction-with-life questionnaire (Diener, Emmons, Larsen and Griffin, 1985) or the PANAS (positive and negative affect schedule) (Watson, Clark and Tellegan, 1988). By measuring our subjective wellbeing, we can make informed decisions about factors that influence our quality of life within a community.

➡️

OVER TO YOU

- When are you most content at work? What are you doing? How are you feeling?
- How satisfied are you with life at the moment? Which factors contribute to this?

Eudaimonic wellbeing

Eudaimonic wellbeing is another concept that originates from ancient Greek philosophy, particularly from the works of Aristotle. It emphasises the pursuit of a meaningful and fulfilling life through focus on personal growth, virtue and realising one's potential. The term eudaimonia can be translated as flourishing or living well.

Unlike hedonic wellbeing, which focuses on seeking pleasure and avoiding pain, eudaimonic wellbeing emphasises the pursuit of a life that is deeply satisfying and meaningful. You may recall that self-actualisation sits at the top of psychologist Abraham Maslow's famous 1943 hierarchy of needs. The birth of humanistic psychology in the 1960s saw renewed attention paid to eudaimonia. Eudaimonic wellbeing recognises that a fulfilling life focuses on long-term flourishing and personal growth. This holistic approach considers an individual's values, relationships, personal development and contribution to society. It requires us to engage in activities that align with our values and strengths.

Eudaimonic wellbeing comes from the knowledge that we are making a difference, which includes noticing the effect we have on others. As Professor John Hattie famously says, 'know thy impact'. Great teachers ask questions, share a common vision and notice visible examples of student learning. When we pay attention to the positive effects of our efforts on student outcomes, we build self-efficacy that strengthens our eudaimonic wellbeing.

OVER TO YOU

- What is most meaningful to you about being an educator?
- Describe a time when you had a positive impact on a student or colleague. What did you do and what effect did this have?

Self-determination theory

We can explore eudaimonic concepts through self-determination theory (SDT), developed by Ryan and Deci (1991). While more a motivation theory than a wellbeing theory, SDT does look at personal achievement for self-efficacy and self-actualisation. At its core, SDT identifies three innate psychological needs that drive choices:

1. **Autonomy.** The need to choose what we will do (being an agent of our own life).
2. **Competence.** The need to feel confident in our skills and ability.
3. **Relatedness.** The need to have human connections that are close and secure, while respecting autonomy and facilitating competence.

When our psychological needs of autonomy, competence and relatedness are met, our motivation and wellbeing are enhanced. Self-determination plays a crucial role in the job satisfaction, engagement and overall wellbeing of teachers. Teachers who are involved in decisions about curriculum design, teaching methods and professional learning feel more engaged. Knowing that they have the knowledge and skills to do their job well makes them feel that their work is meaningful. Opportunities to collaborate and share resources make them feel validated and recognised for their work.

Real-life example

Nick was passionate about teaching, but noticed that he was becoming irritated and stressed by the increasing demands of compliance and administration. He engaged in a professional learning session with me to discover self-determination theory (SDT). In the session we discussed how individuals thrive when they have a sense of autonomy, competence and relatedness in their work. Intrigued, he reflected on his teaching practices and discovered that his autonomy was being undermined by pressure to adhere to a standardised curriculum, leaving little room for him to tailor his teaching. This lack of autonomy made him feel constrained and disconnected from his passion.

Moreover, he realised that he was struggling with feelings of incompetence. He constantly compared himself to other teachers and felt the need to meet unrealistic expectations. The pressure to achieve high test scores and meet stringent performance standards left him feeling inadequate and overwhelmed. He recognised that his sense of connection with his students was also compromised.

Time constraints and an emphasis on academic achievement reduced opportunities to build meaningful relationships, which left him feeling disconnected and unfulfilled.

Through a coaching conversation, Nick and I discussed options for becoming more self-determined. He decided to actively regain his autonomy by finding creative ways to infuse his own teaching style into the curriculum. To enhance his sense of competence, Nick focused on growth and progress rather than perfection. He encouraged his students to set personal goals coupled with celebrations of success. This approach boosted not only the students' confidence, but also helped Nick regain his confidence as an effective teacher. Nick dedicated time to connecting with students by engaging in conversations beyond academics and showing genuine interest in their wellbeing. By embracing the principles of self-determination theory, Nick created a more positive and engaging environment for himself and others, and felt more fulfilled in his role.

Psychological wellbeing

Psychologist Carol Ryff extensively researched this area to create a comprehensive model of the interconnected psychological elements that help us thrive (1989):

- **Self-acceptance.** Accepting your body, emotions, thoughts and past experiences. Having a positive view of yourself, encouraging positive self-talk rather than self-criticism.
- **Environmental mastery.** Adapting and dealing with your circumstances, appreciating the journey, seeing positives in your surroundings, overcoming adversity.
- **Personal growth.** Learning from your experiences and reactions, being open to new experiences and challenges, viewing your actions objectively, finding resources for self-improvement.
- **Purpose in life.** Setting clear and acceptable goals, finding a purpose that will give your life meaning, having direction.
- **Autonomy.** Having choices, being able to make decisions, not being easily influenced by the opinions of others.
- **Positive relationships.** Being able to interact with other people open and sincerely, having bonds with others, maintaining quality relationships, trusting, feeling valued, having empathy.

People with higher psychological wellbeing are more likely to live healthier and longer lives (Kubzansky, Huffman, Boehm and Hernandez, 2018). This is because individuals who exhibit personal growth, environmental mastery and autonomy are more likely to engage in favourable habits such as physical activity, leading to a healthy lifestyle. Psychological wellbeing is also associated with creative thinking, pro-social behaviour and life satisfaction.

(→)

OVER TO YOU

- When you look at your life, how satisfied are you with where you are now?
- How in charge of your life do you feel?
- How willing and energised are you to learn and grow?

Positive psychology and the PERMA(H) pillars

Positive psychology is a humanistic form of psychology generally accepted as the scientific study of what makes like worth living (Peterson, 2008).

Martin Seligman became known as the father of positive psychology after using the term in his 1998 address as President of the American Psychological Association. He called on psychology researchers to move away from a focus on preventing mental illness to consider how we could develop 'courage, hope, interpersonal skills, perseverance, honesty, work ethic, capacity for pleasure, future-mindedness and capacity for insight' (Seligman, 2002).

Falecki and Mann (2020) provide a good definition of teacher wellbeing through the lens of positive psychology:

> *The psychological capacity for teachers to manage normal stressors within the profession, including awareness of positive emotional states. This includes setting authentic goals, celebrating accomplishments, maintaining positive connections with others and reflecting on meaning and impact.*

In his book *Flourish* (2011), Seligman merges theories in eudaimonic wellbeing and subjective wellbeing to identify five areas that when developed could contribute to flourishing. These are typically expressed in a popular model known as the PERMA pillars.

We increase our wellbeing when we:

- Track **positive** emotions
- **Engage** with our strengths
- Form positive **relationships**
- Connect to **meaning** and purpose
- Celebrate **accomplishments**

The pillars of positive emotions and engagement relate to hedonic wellbeing, as they help us feel good. The pillars of meaning and accomplishment relate to eudaimonic wellbeing, as they help us function well or do good work. The pillar of relationships is found in the middle of the PERMA model because it can support both hedonic and eudaimonic wellbeing. In recent years, many psychologists have added health as a pillar in recognition of the strong relationship between mental and physical wellbeing.

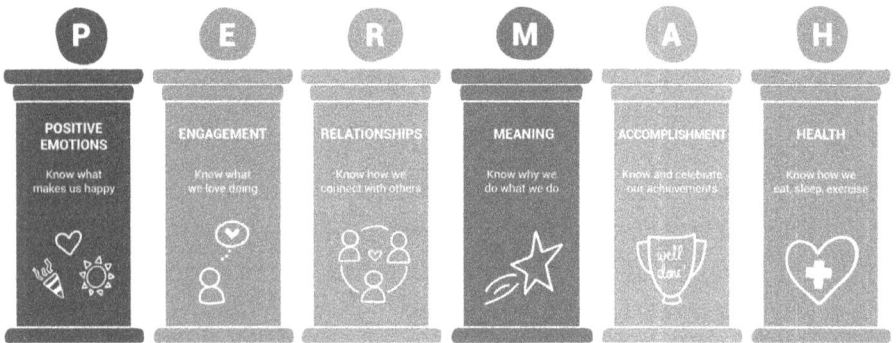

Let's take a closer look at each of the PERMAH pillars and see how they can help us nourish our wellbeing at work.

Positive emotions

Speak to any teacher and they will tell you how they love seeing children grow and learn. Our hearts light up when students have that *aha!* moment, and we experience great joy when our classes immerse themselves in lessons that we have painstakingly prepared.

The challenge is that we have what is called a negativity bias. This is a cognitive phenomenon in which negative events or information have a greater impact than positive ones on our thoughts, emotions and behaviours. It reflects our tendency to notice, remember and be affected by negative experiences, and can lead to a disproportionate focus on threats or challenges. When we have positive and negative experiences daily, the negative ones seem to have more pull than the positive ones.

Think about it: you spend your day ticking off a hundred items on your list and do most of them well, but one little thing doesn't go to plan. What do you ruminate on when you get home? If you are like me and most other people, it will be that one negative experience. This natural phenomenon is our negativity bias at work.

The good news is that we don't have to continue thinking in this way.

Let's look at positive emotions researcher Barbara Fredrikson's broaden and build theory. Fredrickson argues that negative emotions narrow thinking and actions, while positive emotions broaden them (1998). According to the theory, positive emotions broaden our cognitive and emotional resources. They enable us to think more creatively, explore new possibilities and build social connections. Over time, the accumulation of these positive emotions and experiences leads to the development of personal resources such as resilience, psychological wellbeing and improved overall functioning.

Think back to a time when you were in a moment of high pressure, perhaps feeling overwhelmed and worried about meeting the demands in front of you. As you sit at your desk you find that your thinking is blurred, decisions are difficult and mistakes seem frequent.

Emotions can be influenced by the way we explain events to ourselves, otherwise known as our explanatory style. This phenomenon affects our interpretation of experiences and our thinking patterns. We can broadly refer to it as optimistic or pessimistic thinking. Optimists exhibit fundamentally different coping mechanisms to pessimists, especially during turbulent times in organisations (Luthans et al., 2015). When we learn to reflect on our explanatory style, we reframe perspectives that can help us move from a position of perceived helplessness to one of optimism.

It's important to note that all emotions play an important role in our human experience. Focusing on positive emotions doesn't mean ignoring negative emotions. It's simply an acknowledgment that positive emotions bolster us in times of need, helping us be more robust when challenges or negative emotions arise. Identifying and tracking positive emotions is not fluffy New Age rhetoric, but an evidence-based strategy to help strengthen wellbeing.

*E*ngagement

Engagement is a commonly used word in education. It generally means being fully immersed in an intrinsically motivating activity with a balance of challenge and skill, resulting in an experience of 'flow' (Csikszentmihalyi, 1990). Engaged people tend to be curious, passionate and likely to persevere in attaining goals. They feel involved in life, rather than helpless or at the mercy of others.

Recognising behaviours or skills that energise you is a great place to start cultivating engagement. This is where reflection on your character strengths can boost your wellbeing throughout the day. These strengths lie at the foundation of positive psychology, as they give us a common language to describe what is best in human beings.

Character strengths are known as the building blocks of flourishing. They are the positive traits at the core of our identity. These strengths 'are patterns of thinking, feeling or behaving that, when exercised, will excite, engage and energise you, and allow you to perform at your optimum level' (Linley, Willars, Biswas-Diener, 2010). We can use character strengths to increase our engagement.

Character strengths are not skills or talents, but the driving force of who we are. They are unlike talents (what we do naturally), skills (what we train ourselves to do), interests (our passions) or values (what we internally hold dear to us). In the words of Ryan Niemiec, 'talents can be squandered, resources can be quickly lost, interests wane, skills dimmish over time, but when all seems lost, we still have our character strengths' (Niemiec, R. M., p. 17, 2017).

Teachers who use their character strengths experience a higher level of engagement in their work. A study by Harzer and Ruch (2015) found that teachers who identified and used their strengths reported greater job satisfaction, engagement and vitality. By aligning our work with our innate strengths, we can experience authenticity, meaning and accomplishment. Research by Waters, Loton and Jach (2019) revealed that teachers who intentionally applied their strengths in the classroom reported higher levels of enthusiasm and student engagement. Strength-focused teachers are better equipped to navigate challenges, bounce back from setbacks and maintain their wellbeing (Seligman, 2011). This resilience allows them to remain engaged and motivated even in the face of demanding situations, ultimately promoting their long-term engagement and career satisfaction.

Encouraging teachers to develop their strengths positively impacts their professional development and growth. Research by Park, Peterson and Seligman (2006) demonstrated that teachers who received training in identifying and using their strengths experienced increased job satisfaction, commitment and engagement. This highlights the importance of incorporating character strengths into teacher training and professional development programs to promote sustained engagement.

Researchers at the VIA Institute on Character have identified 24 character strengths that are grouped into six categories known as virtues. While we each have the ability to call on all 24 strengths, science has shown that identifying and using our top five 'signature strengths' can increase wellbeing and life satisfaction.

Character strengths and virtues

Virtue: Strength of head (wisdom)		
	Description	Reflecting on strengths
Creativity	I think of interesting ways to solve real-world problems. I enjoy trying new ways to do things.	How will you plan ways to be creative with colour, come up with new ideas or creatively solve problems this week?
Curiosity	I enjoy discovering new places, people and experiences. I am naturally curious and like to know how things work.	What are you curious about learning this week? Who could you ask, to find out more?
Judgement	I am open-minded and like to explore all points of view before making a decision. Facts are important to me.	What decisions this week would benefit from making a list of pros and cons?
Love of learning	I gain a sense of satisfaction from learning new things. I am keen to learn new knowledge and skills that go beyond curiosity.	What have you recently learned from a book, article or website? Who can you share this with?
Perspective	I listen well to others and offer wise counsel. I am able to see the big picture and people look to me for advice.	Who could you offer support or advice to this week?

Virtue: Strength of heart (courage)		
	Description	**Reflecting on strengths**
Bravery	I speak up for what is right even if there is opposition. I act on my convictions even in difficult situations.	How can you be brave this week by doing something new or challenging?
Perseverance	I finish what I start without letting obstacles distract me. I take pleasure in completing tasks that I start.	What is a goal or a project you would like to complete sometime soon? When will you have this done?
Honesty	I speak the truth and take responsibility for my own feelings and actions. I act with integrity and authenticity, without pretence.	When might you need to have an honest conversation with someone this week?
Zest	I approach life and tasks with excitement and energy. I seek out new adventures and don't do things half-heartedly. I feel alive when seeking out new adventures.	What is an activity that excites and energises you? When will you plan to do this?

Virtue: Strength of giving (humanity)		
	Description	**Reflecting on strengths**
Love	I value close relations with others, particularly when sharing and caring are reciprocated. Supporting other people's happiness is as important as achieving my own.	Who is someone you truly appreciate? How and when can you let them know?
Kindness	I believe everyone should be treated equally. I treat others with compassion. I enjoy doing favours and good deeds for others.	What is kind gesture you could do this week? Who will you be kind to and when?
Social intelligence	I am aware of the feelings of others when they communicate with me. I understand what motivates others including how they may interpret my actions.	How could you show others you care? Who will you speak to and when?

Virtue: Strength of teamwork (justice)		
	Description	**Reflecting on strengths**
Teamwork	I work well with others. It's important for me to help others who may need it. I am loyal and productive in teams, always doing my fair share.	What team project could you be working on this week? How will you be a good team player?
Fairness	I believe everyone should be treated fairly and with respect. I give everyone a fair chance regardless of who they are or where they come from.	How could you ensure equity and fairness in something at work this week? What will you do?
Leadership	I am able to organise others to get things done. I not only plan the course of action but motivate and encourage others in moving towards achieving a common goal.	I am able to organise others to get things done. I not only plan the course of action but motivate and encourage others in moving towards achieving a common goal.

Virtue: Strength of self (temperance)		
	Description	**Reflecting on strengths**
Forgiveness	It's important for me to forgive others and repair relationships. I forgive others when they hurt my feelings, giving them a second chance.	How you could show compassion and empathy to someone at work this week?
Humility	I let my accomplishments speak for themselves rather than bragging about them to others. I am open to feedback and am aware of my shortcomings.	How could you listen more closely to others so they feel valued?
Prudence	I think carefully before acting and avoid taking undue risks. I am cautious in my approach to decision-making and prefer to take my time than rush into things.	What is something that needs careful consideration at work this week? How will you do this?
Self-regulation	I am self-disciplined and can stick to specific plans when I set them for myself. I am able to manage my impulses.	What is a habit or behaviour that needs regulating at work this week?

Virtue: Strength of spirit (transcendence)		
	Description	**Reflecting on strengths**
Appreciation of beauty and excellence	I notice the level of skills used to achieve excellence. I appreciate skilled performances and admire craftmanship, whether this be in a work setting, art, mathematics or in nature.	What is a something at work that you can make beautiful or will appreciate the beauty of?
Gratitude	I take time to thank people and am grateful when good things happen to me. Even when things don't go my way, I find a way to notice the positives in my life.	How will you practice gratitude at work this week?
Hope	I expect good things to happen in my life and prefer to look on the bright side. I set goals with actions to achieve these knowing that despite challenges there is always hope.	What is a goal you are wanting to achieve? What actions will get you there?
Humour	I add laughter and humour to situations. If someone is upset, I will try to make light of the situation to cheer them up. People say I'm fun to be around.	How will you plan ways to play more at work this week?
Spirituality	I believe every life has a purpose that connects to a higher power. Knowing I am being meaningful in what I do is important to me.	What is most meaningful to you this week? Why?

OVER TO YOU

Take a few minutes to complete the Values in Action Character Strengths Survey at www.viacharacter.org. This is a free self-report measure that gives a rank order of the 24 character strengths.

- What are your top five character strengths?
- Using the preceding table of character strengths, answer the questions that match your top five strengths

Relationships

Relationships are a basic psychological need and foster a sense of belonging. Just like food and air, we need social relationships to thrive (Diener and Biswas-Diener, 2008). Positive relationships help develop our core internal resources, including social and emotional skills (Roffey, 2012). As Seligman states, 'other people are the best antidote to the downs in life and the single most reliable up' (2011). Positive psychology's focus is not solely on improving personal wellbeing, but also on how to better connect and contribute to the lives of others.

Positive relationships with colleagues have been linked to increased job satisfaction and reduced burnout among teachers (Schwab, 2018). Collaborative and supportive interactions foster a sense of camaraderie that allows teachers to feel valued, understood and emotionally supported (Hakanen, Bakker and Schaufeli, 2006). This, in turn, reduces stress and promotes a positive work environment conducive to personal and professional growth (Taris, Schaufeli and Verhoeven, 2005).

Strong connections with students enhance teacher wellbeing. Positive teacher-student relationships contribute to increased job satisfaction, motivation and sense of purpose (Roorda et al., 2011). When teachers establish caring and respectful connections with their students, they experience greater job engagement and a sense of accomplishment (Oberle and Schonert-Reichl, 2016). Moreover, positive student-teacher relationships create a supportive classroom climate that fosters learning, cooperation and emotional wellbeing for students and teacher alike (Jennings and Greenberg, 2009).

Let's not forget that school leaders and parents also positively affect teacher wellbeing. When teachers feel supported and valued by their superiors, they experience greater job satisfaction and reduced stress levels (Van Droogenbroeck, Spruyt and Vanroelen, 2014). Collaborative partnerships with parents, characterised by open communication and common goals, promote a sense of shared responsibility for student success and contribute to teacher wellbeing (Jennings and Greenberg, 2009).

Positive relationships form an important part of what is known as social capital, or the level of trust and shared values expressed by a group. The more we invest in promoting positive relationships at work, the more we enhance the resources within ourselves and the people around us. We can do this by sharing ideas, advice, emotional support and opportunities to grow. Investing in the development of social capital creates conditions where people feel trusted and safe with a sense of belonging. People with high levels of social capital are calm, kind and supportive.

When we joyfully connect with another person, oxytocin released into the bloodstream reduces stress and anxiety. Even a micro-moment of genuine connection can spark an upwards spiral of care between people, known as 'positivity resonance' (Fredrikson, 2013).

Professor Jane Dutton from the University of Michigan explains how short positive interactions with people can form high-quality connections. These are interactions that leave you feeling energised and invigorated. High-quality connections occur not only with people to whom you are already deeply connected, but also in fleeting moments.

On the other hand, corrosive connections sap energy. We become de-energised trying to protect ourselves from threats such devaluation, disrespect and distrust. When we are fearful of how others will respond to us at work, our capability and motivation to function are reduced. We become so focused on managing the negativity that comes from the connection that we lose sight of the goal at hand.

Teaching is a collaboration that requires honest and trusting relationships free of judgment yet stable enough to manage challenging conversations. Positive relationships are needed to facilitate conversations between each layer within the system: communicating with students, sharing feedback with parents, working together with colleagues. Being able to articulate our personal needs is a way of practicing self-care, and seeking support requires the understanding that a confidant will be trustworthy.

OVER TO YOU

- Who are two people who support you at work? How do they offer support?
- Who are two people whom you support at work? How do you offer support?

Meaning

Meaning is our sense of where we fit in the world. Meaning is created by our understanding and acceptance of ourselves, the world around us and our place among people (Steger, 2008). People who feel they live meaningful lives have greater longevity, life satisfaction and greater overall wellbeing (Bonebright et al., 2000). Those who feel their lives are lacking in meaning are at risk of depression, poor social relationships and substance abuse (Kim, 2017).

According to the PERMAH pillars, meaning refers to the intrinsic value and joy we feel when contributing to society. It is strongly linked to a sense of purpose, efficacy and self-worth (Baumeister and Vohs, 2005). Finding meaning at work can increase wellbeing and decrease hostility, stress and depression (Steger et al., 2006). Engagement in meaningful work can increase commitment, connection, happiness, satisfaction and fulfilment (Wrzesniewski et al., 2013).

Finding meaning in work can profoundly impact our sense of purpose, job satisfaction and overall wellbeing. When teachers perceive their work as meaningful, they are more likely to feel motivated and engaged in their daily activities (Steger, Dik and Duffy, 2012). Teachers who feel a sense of meaning understand the broader impact of their role in shaping the lives of their students and contributing to society, which can fuel their dedication and commitment to their teaching responsibilities (Cohen, 2006). This sense of purpose can act as a source of inspiration and resilience, helping them to navigate challenges and maintain a positive outlook.

Meaning in teaching is closely tied to personal values and beliefs. When teachers align their work with their core values and principles, they experience a sense of coherence and authenticity (Norrish, Williams, O'Connor and Robinson, 2013). These teachers feel that their work is congruent with their beliefs about education, child development and making a positive difference in the lives of their students. This alignment fosters a deep sense of fulfillment and personal satisfaction, contributing to overall wellbeing (Duffrin, 2009).

By fostering meaningful relationships with students, teachers create an environment where learning is not just about academic content but also about personal growth and development (Schutz and Zembylas, 2009). These connections enable teachers to witness the progress and successes of their students, reinforcing the significance of their role and providing a sense of fulfillment (Wong, 2014). Furthermore, collaboration with colleagues and a sense of belonging within the school community contribute to a shared sense of purpose and create opportunities for collective meaning-making (Yin, Huang and Wang, 2020).

Inspirational speaker Simon Sinek famously asks us to 'start with why'. He suggests that knowing why we do something is more powerful in building momentum for action that knowing what to do (Sinek, 2009). By connecting to our 'why', we are connecting to meaning. Living a meaningful life helps us align with something bigger than ourselves and can contribute to a sense of belonging (Seligman, 2011). In Chapter 9 you will have a chance to create your own 'why' statement.

➡️

OVER TO YOU

- Why did you decide to be an educator?
- What is most meaningful to you in your work?

_A_ccomplishment

Accomplishment refers to the application of personal skills and effort as we move towards a desired goal (Seligman, 2011). A significant factor in accomplishment is the clarity of a goal. Goals direct attention and effort, and can lead to new knowledge and skills. They fuel our intrinsic motivation, making us more committed, driven and enthusiastic about our work (Locke and Latham, 2019). The more SMART (specific, measurable, assignable, realistic and time-related) the goal, the more likely its achievement.

Luckily for us, schools are achievement arenas for both students and staff. When teachers achieve goals, meet academic targets and witness the progress and success of students, it brings a sense of fulfillment and pride (Skaalvik and Skaalvik, 2018). Accomplishment provides tangible evidence of our effectiveness as educators, reinforcing our belief in our professional competence and enhancing job satisfaction. The pursuit of accomplishments stimulates our passion for teaching and encourages us to invest our energy in our instructional practices.

A sense of accomplishment contributes to teacher self-efficacy and confidence. When teachers accomplish challenging tasks or witness the positive impact of their instructional strategies, it strengthens their belief in their ability to make a difference to students (Tschannen-Moran and Hoy, 2001). This, in turn, boosts self-confidence and enhances overall wellbeing and professional growth.

Accomplishment supports teacher resilience and perseverance. Teachers who overcome obstacles, address difficulties and achieve positive outcomes build their resilience and equip themselves with the tools to navigate future challenges (Martin et al., 2013). Accomplishment serves as a source of motivation and resilience, helping teachers to recover from setbacks and maintain their wellbeing in the face of adversity.

Psychologist Carol Dweck suggests that more important than our goals is our belief in what is possible. This is where growth mindset thinking can help us learn and achieve. When we believe in our ability to succeed, we are more likely to persevere and take on hard challenges. We know this is the case for students, because I'm sure that most of us have bought the books and been to the workshops on growth mindsets! But are you yourself engaging in this practice? As educators, we set actionable goals that we track as outcomes. We put a lot of time and effort into these goals, yet we rob ourselves of acknowledging our daily achievements. If we are to flourish, we would benefit from recognising and affirming our achievements regularly.

Health

Physical health is an important part of wellbeing. The health of our body affects the health of our mind. Good nutrition, exercise and sleep habits support the healthy functioning of our nervous system, digestive system and cognitive ability. The better we move, eat and sleep, the better our body is able to restore energy. Through exercise we build stamina, endurance, strength and flexibility. Through nutrition we repair cells with protein, restore energy with carbohydrates and replenish with vitamins and minerals. Through sleep we allow our nervous system to recharge and our body to deeply rest.

Physical health practices contribute to reduced stress levels, increased resilience and enhanced job satisfaction among teachers (Bennett et al., 2020). When teachers feel physically well, they are better equipped to handle the demands of their profession and maintain a positive work-life balance.

Physical and mental health are intertwined, and addressing both aspects is essential for overall wellbeing (Keyes, 2009). Teachers who prioritise their health holistically experience greater job satisfaction, engagement and resilience.

Sleep is one of the most important things we can do to look after ourselves. Unfortunately, recovery is often neglected as a trade-off to performance. The challenge for teachers is being able to turn down the activity of our minds. We typically go through several sleep cycles that include rapid eye-movement (REM) sleep—important for brain development—and Non-REM sleep—important for healing.

Adequate sleep is essential for emotional regulation and overall wellbeing. Inadequate sleep has been linked to obesity, inflammation, aches and pains, and increased risk of Type 2 diabetes. Those who sleep fewer than seven hours per night increase their risk of putting on weight by over 40 per cent. Sleep deprivation has been associated with increased stress levels, reduced cognitive performance and negative mood. A study by Pilcher and Walters (1997) demonstrated that sleep deprivation among teachers negatively impacted their mood, job satisfaction and job performance. Sleep also impacts our capacity to relate well to others. If we don't sleep well, we can be grouchy, irritable and generally unpleasant.

A balanced diet is also important for good health. Food is energy, and food choices can leave us feeling either energised or drained. Teachers often eat on the run and look to easy-to-grab foods. The problem is that these are often jam-packed with sugar, salt and saturated fat. A better diet will have a positive impact on your mood, energy and thinking capacity.

Eating a balanced diet has been linked to improved physical and mental health outcomes. A 2011 study by Jacka et al. found that a diet rich in fruits, vegetables, whole grains and lean proteins was associated with a lower risk of depression. Moreover, a systematic review by O'Connor et al. (2019) revealed that proper nutrition positively influenced cognitive functioning, mental health and overall wellbeing.

Exercise is essential for physical and mental health, and is an effective anti-anxiety treatment. It relieves tension and stress, and boosts physical and mental energy through the release of endorphins. Exercise increases alertness, concentration and cognitive performance, as it transforms the nervous system from fight-or-flight mode to a more balanced response (McGonigal, K., 2020). A meta-analysis by Schuch et al. (2016) demonstrated that exercise interventions had a significant positive effect on reducing symptoms of depression among adults.

OVER TO YOU

Take a moment to complete the free PERMAH profiler survey at www.permah survey.com (Kern and McQuaid, 2019) and answer the following questions.

- Which pillar is your strongest? What actions do you take to support this pillar?
- Which pillar is your weakest? What actions could you take to support this pillar?
- Which pillar would you like to focus on for the next week?
- What are three actions that you could take to strengthen this pillar?

Conclusion

The concept of wellbeing is complex, with many theories to help us understand it. If we are to support the wellbeing of teachers, we need to build our knowledge of these theories. We must explore the related strategies and draw on science to ensure that what we are doing is meaningful and appropriate. This can help us transform our individual and collective wellbeing at work.

Social and emotional competence

The complexities of teaching

Teaching is an emotional vocation. It is likely that a typical day will result in a melting-pot of emotions, with positive and negative bubbling to the surface at any one time. We are expected to regulate not only our own emotions, but those of our students as well. The skills to identify, understand, use and manage these emotions require emotional intelligence. This becomes the foundation for social and emotional competence.

Social and emotional competence plays a central role in determining how, what, when and why we do what we do in the classroom. When we display social and emotional competence, we are better able to create an environment that is positive, supportive and well-organised. We are also able to better connect and collaborate with colleagues and address parental needs.

Social and emotional competencies are critical for teachers to avoid burnout and strengthen wellbeing. Regulating our emotions before reacting to student misbehaviour, unwinding in healthy ways after a busy day and identifying our internal motivations are all ways of using emotional intelligence to feel better and function well. Research shows that effective mastery of social and emotional competencies is associated with greater wellbeing and better school performance, while a failure to achieve competence in these areas can lead to a variety of personal, social and academic difficulties (Greenberg et al., 2003). Mindfulness can be an effective strategy to help us to notice and regulate emotions as they arise.

Building social and emotional competence (SEC)

Teachers with strong social and emotional skills are better able to create supportive learning environments, form positive relationships with their students, promote their own wellbeing and enhance student learning. The Collaborative for Academic, Social and Emotional Learning (CASEL) is a leading organisation dedicated to promoting social and emotional learning (SEL) in schools. CASEL describes SEL as 'the process through which children and adults understand and manage emotions, set and achieve positive goals, feel and show empathy for others, establish and maintain positive relationships, and make responsible decisions' (2023).

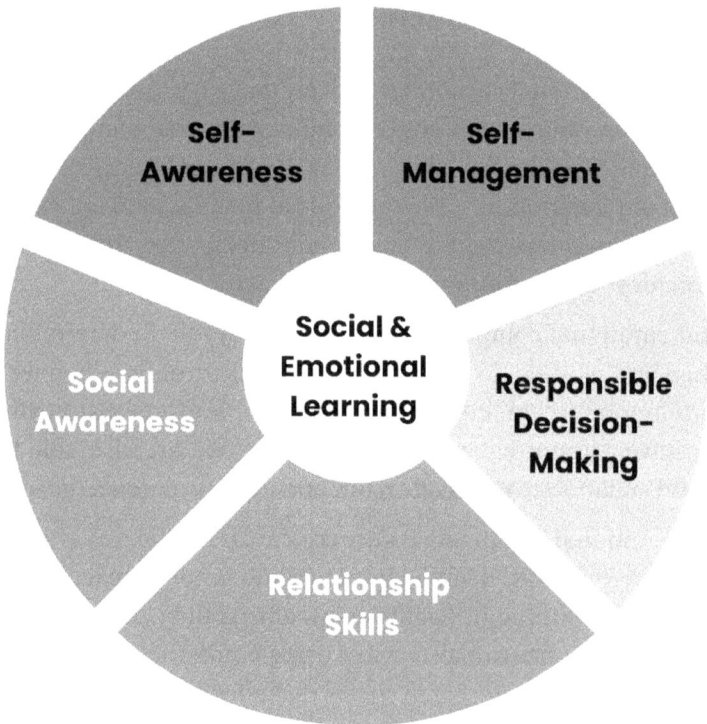

According to CASEL, classrooms led by teachers with strong SEL skills are more cooperative, respectful and academically productive. Socially and emotionally competent teachers are better able to form positive relationships with their students. According to a meta-analysis of 213 studies conducted by CASEL, students who had teachers with strong SEL skills had more positive attitudes towards school, better behaviour and higher academic achievement (Greenberg et al., 2016). Teachers with strong SEL skills are better able to promote students' social and emotional development, which in turn leads to improved academic outcomes.

CASEL's research indicates that teachers who model social and emotional competencies are more effective at teaching SEL to students. Teachers trained to implement a SEL curriculum report greater efficacy in managing student behaviour, higher levels of personal accomplishment, lower job-related anxiety and depression, higher-quality classroom interactions, greater personal engagement and greater perceived job control (Greenberg et al., 2016). The training of teachers to deliver student SEL programs may serve to build resilience not only in students but in the teachers themselves.

CASEL's model includes the development of five interrelated competencies that form a SEL wheel. If we as educators want the skills to better manage and plan for our wellbeing, we need to learn and build on these five competencies.

Real-life example

Lee had always been enthusiastic about educating young minds, but lately her enthusiasm had dimmed. The classroom had become a battleground of emotions. She found herself struggling to manage her feelings at work, fluctuating between angry outbursts and tears. The pressures of dealing with diverse personalities, parent-teacher communication and a changing educational landscape had taken their toll. She felt overwhelmed, socially disconnected and on the brink of burnout.

Lee engaged me as a coach to support her through the debilitating feelings of stress and overwhelm she was experiencing. We began by exploring her emotions to help her become more aware of her reactions and triggers. Through deep reflection and specific exercises, Lee was able to identify her emotions and manage them in ways that did not negatively affect anyone. Slowly but surely, she started to become more attuned to her feelings and needs. Her relationships with staff and students improved as she began to gauge not only her own emotions but those of the people around her.

OVER TO YOU

- In your own words, what does it mean to be emotionally competent?
- What does it mean to be socially competent?

Self-awareness

Self-awareness is the ability to accurately recognise our emotions and thoughts and their influence on our behaviour. Self-aware people accurately assess their strengths and limitations, act with confidence and purpose, and ensure that their values link to their identity with integrity. They are also capable of examining potential biases to question their assumptions and patterns.

Teachers who are socially and emotionally competent demonstrate self-awareness by understanding their emotions, values and strengths. They recognise how these impact their teaching practices and are constantly checking in with themselves. They are reflective and open to feedback, able to regulate their emotions in the classroom.

OVER TO YOU

- How do you practice self-awareness?
- What qualities do you admire in people whom you believe to be self-aware?

Self-management

Self-management is the ability to regulate our emotions, thoughts and behaviours effectively in different situations. It means managing stress, controlling impulses, motivating ourselves when needed and working to achieve goals. Self-management gives us agency to plan for personal and collective endeavours.

Teachers who are socially and emotionally competent demonstrate self-management by being able to regulate their emotions and behaviours, set goals and manage stress effectively. They stay focused and maintain a positive attitude in the face of challenges.

→

OVER TO YOU

- How do you manage your emotions when you feel overwhelmed of stressed?
- How do you respond when you are unable to manage your emotions effectively?

Social awareness

Social awareness is the ability to see the perspective of others and empathise with them. People with social awareness demonstrate compassion and understand needs that they may not share. They see strengths and can express gratitude when needed. They recognise social norms and the influence of these across organisations and communities.

Teachers who are socially and emotionally competent create a safe and inclusive classroom environment and are able to respond appropriately to the needs of individual students.

→

OVER TO YOU

- How aware are you of the people around you when you are in a group?
- Do you notice body language, tone and mood?

Relationship skills

Teachers must have the ability to establish and maintain healthy and rewarding relationships with diverse individuals and groups. This means communicating well, actively listening, cooperating, resisting inappropriate social pressure, negotiating conflict constructively and seeking and offering help when needed.

Teachers who are socially and emotionally competent demonstrate strong relationship skills by building positive connections with students and their families, communicating effectively and resolving conflicts in a constructive way.

➡️

OVER TO YOU

- What skills help you develop positive relationships?
- How do you respond or manage difficult relationships?

Responsible decision-making

This is the ability to make caring and constructive choices about personal behaviour and social interactions based on consideration of ethical standards, safety concerns and social norms. Responsible decision-makers demonstrate curiosity and open-mindedness when seeking solutions and recognise the significance of critical thinking when evaluating consequences.

Teachers who are socially and emotionally competent demonstrate responsible decision-making by making thoughtful and ethical choices that take into account the wellbeing of their students and the broader community.

➡️

OVER TO YOU

- Recall a difficult but successful conversation you've had with someone. What did they say? How did you choose to respond?
- Which responsible choices did you make that contributed to the outcome?

Social and emotional competence builds resilience

Resilience is not about avoiding or eliminating stress, but about developing the skills and resources to effectively manage and respond to it. The concept of resilience in education is not new, featuring strongly in student wellbeing programs. When it comes to teacher resilience, some researchers highlight the importance of limiting risk factors such as workload and overcommitment while others look at building protective factors such as social and emotional competence.

Leading expert Ann Masten defines resilience as 'the capacity of a dynamic system to adapt successfully to disturbances that threaten system function, viability or development' (2014). Based on this definition, resilience does not mean withstanding risk but rather changing to accommodate it. This requires us to maintain positive functioning despite setbacks or demanding circumstances, much like a tree on the side of a cliff will adapt its shape to suit the environment.

Teachers are expected to thrive in chaos, while we admire people who learn and grow in tough times, yet spend little effort building the internal resources to achieve this. Resilience is crucial for teachers to effectively manage the inherent stressors and demands of their work.

Benefits of resilience

Resilient teachers demonstrate flexibility and adaptability in their teaching practices and approaches. They adjust their strategies based on student needs, changing circumstances and evolving educational contexts. They are adept at identifying challenges, finding creative solutions and effectively managing classroom issues. They exhibit strong problem-solving and decision-making abilities.

Resilient teachers maintain a positive outlook even in challenging situations. With a strong sense of self-awareness, they focus on strengths, possibilities and potential solutions rather than dwelling on problems. They actively seek and use social support networks such as colleagues, mentors and professional communities. They engage in collaborative problem-solving, share experiences and draw strength from their relationships.

Resilient teachers prioritise self-care and are better equipped to manage stress. They engage in activities that promote physical and mental health, maintain work-life balance and seek resources for personal rejuvenation. Resilient teachers are more likely to experience reduced burnout, higher levels of engagement and improved student outcomes.

How can teachers learn social and emotional skills?

Teaching social and emotional skills to teachers requires thoughtful planning and implementation. High-quality professional development is essential. Programs that are focused on SEL can be effective in building teachers' knowledge and skills in this area (Durlak et al., 2011). These programs should be evidence-based, interactive and provide teachers with opportunities to grow in self-awareness and emotional management.

Schools can create a culture of SEL by modelling these skills in professional interactions such as staff meetings or conversations between teachers and administrators (Jones and Bouffard, 2012). This can reinforce the importance of positive relationships and create a shared language for discussing emotional responses to situations.

SEL should be integrated into curriculum and instruction, rather than being taught as a separate program (Durlak et al., 2011). This means that it should be used in everyday language within a school community. Teachers can thus be assured that students are receiving consistent messages about social and emotional skills across all subjects, reinforcing the importance of these skills.

Teachers need ongoing support to effectively grow as professionals and to implement SEL in their classrooms (Brackett and Katulak, 2006). Useful opportunities include coaching sessions, peer support networks and professional learning communities.

Challenges of teaching social and emotional skills to teachers

There are several common barriers that can hinder teachers' learning of social and emotional skills. The most significant is limited access to comprehensive training and professional development programs focused on social and emotional skills (Jones and Bouffard, 2012). This lack of training not only impacts coping capacity, but can hinder teachers' ability to acquire the knowledge and skills needed to teach SEL to students.

It's no surprise that time constraints due to heavy workloads and competing demands are another barrier (Brackett et al., 2019). Finding time to dedicate ourselves to learning and practicing social and emotional skills can be challenging. This is exacerbated by a lack of institutional support in resources, policies and leadership endorsement (Greenberg et al., 2017). When SEL is not consistently prioritised or emphasised at a school or regional level, it can be challenging for teachers to commit to learning and implementing social and emotional skills (Jones and Bouffard, 2012). Inconsistent messaging or competing priorities may create confusion or a lack of clarity regarding the importance of SEL.

One globally recognised evidence-based program that aims to promote the social and emotional wellbeing of educators and students is the Cultivating Awareness and Resilience in Education (CARE) program. This four-day training program is presented across several months and addresses three key areas: building emotional literacy, teaching mindfulness-based interventions, and promoting empathy and compassion through listening and caring practices. Building Resilience in Teacher Education (BRITE) is another program shown to improve teacher wellbeing by reducing stress, anxiety and burnout (Suldo et al., 2015). It also increases resilience among pre-service teachers, leading to greater psychological wellbeing and reduced stress levels (Singleton, Kruse and Lichtenberg, 2020).

Some teachers may be resistant to change or sceptical about the value and relevance of social and emotional skills in the classroom (Schonert-Reichl et al., 2017). They may feel that focus should solely be on academic subjects, or be hesitant to learn new approaches that emphasise SEL. This resistance can be related to concerns about time and resources, or a lack of understanding about the benefits of these approaches.

If we want our educators to think, feel and act differently, we must give them the time and resources to learn new skills. Just as we prioritise the social and emotional development of students, so must we prioritise that of teachers.

Mindfulness, a tool to build SEL

Mindfulness is a form of meditation that involves being fully present and engaged in the present moment with a non-judgmental and accepting attitude. Encompassing self-awareness, self-regulation, empathy and interpersonal effectiveness, it is a common practice for the development of social and emotional competence.

Mindfulness can help us regulate our emotions by increasing awareness of our internal experience without reactivity or judgment. This can help individuals recognise their emotions and respond to them adaptively, rather than being overwhelmed by them (Hölzel et al., 2011). Mindfulness has been found to be effective in reducing stress and burnout among teachers. In a study of 78 primary school teachers, those who participated in an eight-week mindfulness-based program to reduce stress reported significant reductions in perceived stress and burnout (Flook et al., 2013).

Mindfulness has also been shown to regulate the nervous system activating the parasympathetic nervous system (PNS) and reducing the activation of the sympathetic nervous system (SNS). The PNS is responsible for the body's relaxation response, while the SNS is associated with the fight-or-flight response. Focusing attention on the breath or body sensations has been found to stimulate the PNS. This activation leads to a decrease in heart rate, blood pressure and respiration rate, promoting relaxation and a sense of calm (Tang et al., 2019). PNS activation also supports restorative processes and aids in recovery from stress. By regulating the nervous system, mindfulness practice helps us shift from a state of reactivity and stress to one of relaxation, resilience and emotional balance. These effects contribute to improved wellbeing, emotional regulation and mental health.

Mindfulness practices encourage self-care and self-compassion, helping teachers cultivate a healthier balance between their professional responsibilities and their personal wellbeing. By nurturing their mental and emotional health, teachers can better support the needs of their students (Roeser et al., 2013). Teachers who practice mindfulness are more likely to create a safe and supportive learning environment that enhances student engagement, reduces disruptive behaviour and improves academic outcomes (Jennings et al., 2013). By positively influencing the classroom climate, we can foster a sense of calm, presence and empathy.

Real-life example

Stephen worked at a challenging secondary school with many students who came from disadvantaged backgrounds. He loved his job, but found it difficult to manage his emotions in the classroom at times. He would often feel overwhelmed, frustrated and angry with his students, and this made it hard for him to connect with them and build a positive classroom culture.

Stephen had learned about the value of mindfulness during a training session and decided to give it a go. He started by incorporating a few minutes of meditation into his daily routine using an app on his phone. Before class, he would sit quietly for two minutes and focus on his breath while he was guided through a mindfulness practice. As time went by, he found himself being conscious of his breathing in class too, where he would take a deep breath and pause before reacting to a difficult situation, giving himself time to think and respond with patience and kindness.

In my conversations with him a few months later, he said that while he still felt the full spectrum of emotions, he was able to catch himself before he reacted. He found that he was being more patient with his students, even when they were being challenging. He was able to stay calm and focused during stressful situations and respond in a way that was more helpful and constructive.

Conclusion

Teachers who are socially and emotionally competent have an awareness of themselves as human beings with the skills to manage their thoughts, feelings and responses. They prioritise self-care, recognising that their own wellbeing is essential to their ability to support their students. They have the ability to regulate their emotions using techniques such as deep breathing and mindfulness.

By cultivating self-awareness, we can identify our needs and take steps to address them. When we are socially and emotionally competent, we also find ourselves better equipped to identify and respond to the needs of others. Socially and emotionally competent teachers take the time to get to know their students as individuals and show them empathy, kindness and respect. They use positive behavioural management strategies, with clear expectations and logical consequences taking the place of punitive measures. They create a positive classroom culture by modelling gratitude, kindness and a growth mindset. They also create opportunities for students to collaborate, communicate and connect with one another.

Psychological health and safety

The effects of our environment

Mental health is a growing topic in education, with millions of dollars allocated to student wellbeing every year. Given the WHO's prediction of depression being the world's leading illness by 2030, we need to consider how we support the psychological health and safety of teachers.

It is estimated that poor mental health costs the Australian economy approximately $11 billion annually due to absenteeism, presenteeism (unproductive time at work) and compensation claims (Beyond Blue, 2014). PwC found that companies that invested in the mental health of their employees saw a positive return of investment of $2.30 for every dollar spent due to productivity gains (2014).

Many psychological wellbeing initiatives for teachers rely on individuals to initiate changes in their behaviour, while the work-related processes that create stress remain unchanged. Although personal wellbeing initiatives such as mindfulness programs and other positive psychology interventions are beneficial, organisations have a responsibility to prevent harm by fostering healthy workplaces. This requires them to establish policies, practices and resources that build employee health, wellbeing and engagement. These include opportunities for professional growth, work-life balance initiatives and health promotion programs.

An organisation's success relies on the wellbeing and productivity of its employees. Recognising this interdependence highlights the need for collaboration between individuals and organisations to create a positive workplace culture. While this book does not cover the implementation of organisational change, it is important that we discuss the impact of systemic decisions on our wellbeing. This chapter aims to provide a comprehensive understanding of the current requirements for organisations, identification of common psychosocial hazards and suggestions for redesigning work to limit stress.

By incorporating strategies such as feedback and recognition processes, workload management, professional development opportunities and supportive leadership, schools can encourage the wellbeing of teachers while maintaining educational excellence. As individuals we can be proactive in thinking about how we contribute to our environment, the support structures we may have available to us but do not access, the personal decisions we make that affect how we work, and what it means to be valued and recognised.

Organisational psychology

The field of organisational psychology applies psychological principles and research methods to understand and improve human behaviour in organisations. Organisational psychologists study how employee wellbeing is affected by the design of jobs. They explore aspects such as workload, autonomy, skill variety and feedback to recommend changes that can make work more engaging and meaningful. Organisational psychologists investigate work design to help organisations create supportive environments that allow employees to effectively manage their work and personal responsibilities. They also analyse factors such as communication patterns, leadership styles and organisational values to promote a positive and supportive culture that fosters employee engagement, satisfaction and wellbeing.

Organisational psychologists can assess organisational culture and identify areas for improvement in schools. Their work with teachers clarifies roles and responsibilities in a way that promotes a healthy work-life balance. They identify strategies for workload reduction, task allocation and time management to prevent stress and burnout. The programs that they develop and implement can reduce stress and build resilience. Educational leaders and administrators may be required to enhance their skills and promote supportive and empowering leadership practices. Supportive systems and policies may also be established to promote a positive work environment.

Common leadership mistakes

It is often assumed that the responsibility for everyone's wellbeing lies solely on the shoulders of those in charge. The role of leaders is, in fact, to foster an environment conducive to growth and provide opportunities for individuals to make choices. We all know the saying: you can take a horse to water, but you can't make it drink.

Leaders must recognise the complexity of individuals. They should draw upon the principles of adult learning (andragogy) and invite staff to be active contributors in finding solutions. Wellbeing initiatives must be launched with caution to ensure that they address genuine concerns. One of the challenges faced by leaders is the lack of time available for strategic thinking about wellbeing. Numerous demands often leave them stretched thin, making it difficult to formulate and implement meaningful plans. Engaging the expertise of professionals can help streamline this process and save valuable time.

Furthermore, leaders often lack familiarity with the field of organisational psychology. While well-versed in educational psychology, they may not possess the specialised knowledge needed to navigate the intricacies of workplace dynamics.

Workplace health and safety

Health and safety is not a new concept. I'm sure we're all familiar with the art of putting yellow lines on stairs to highlight the danger of tripping and laminating hot-water signs in the kitchen to prevent people from scalding themselves. What many of us don't know is that health and safety is a multidisciplinary field concerned with protecting both physical and psychological wellbeing at work.

Global health and safety requirements have been agreed upon by 70 countries belonging to the International Organisation for Standardization. *ISO 45001* provided an initial framework for organisations to establish, implement, maintain and continually improve their occupational health and safety practices.

More recently, *ISO 45003* (2021) specifically addresses psychological health and safety in the workplace. These standards are endorsed by the national guidelines from Safe Work Australia (2019) identifying the requirements of organisations to eliminate or minimise the risk of psychological injuries caused by work.

Psychosocial hazards for teachers

Harm prevention requires a systemic approach that addresses the impact of psychosocial hazards in the workplace. These hazards include:

- High job demand
- Low job control
- Inadequate training and support
- Low role clarity
- Lack of positive feedback
- Inconsistent application of policy and procedures
- Vicarious trauma

These hazards frequently appear in the literature associated with teacher stress and burnout, with high job demand one of the critical factors in teacher attrition (Ed Support UK, 2019).

Workload and time pressure

Teachers have numerous responsibilities: lesson planning, grading, preparing materials, administrative tasks. The demands of teaching multiple classes and maintaining quality standards within limited time frames can lead to stress, exhaustion and overwhelm. Excessive workloads may result in teachers sacrificing personal time.

Emotional demands

Teaching involves emotional labour. Every day we interact with students and respond to their needs and behaviours. These emotional demands can be draining, especially when we have limited resources or support to address challenges effectively. Constant exposure to students' emotional struggles and pressures can also impact our wellbeing.

Classroom management challenges

The need to maintain order and address disruptive behaviours in the classroom introduces significant psychosocial hazards. Teachers may face difficulties in managing large class sizes, handling student conflicts and ensuring an inclusive and respectful learning environment. Dealing with daily classroom challenges can generate stress and frustration, leading to a negative impact on psychological health and job satisfaction.

Lack of support and resources

Inadequate access to materials, technology and teaching aids may impede our ability to deliver quality education and meet student needs. Limited support from colleagues or administrators, inadequate professional development opportunities, a lack of mentoring or guidance, and little recognition and appreciation all negatively impact motivation and wellbeing.

Professional relationships and collegiality

Interactions with colleagues and administrators play a crucial role in our work environment. Poor relationships, a lack of teamwork, conflict and unsupportive organisational culture can create psychosocial hazards. A toxic work environment can result in increased stress, reduced job satisfaction and feelings of isolation. Positive relationships, on the other hand, foster collaboration, support and a sense of belonging.

Role ambiguity and role conflict

Teachers often face role ambiguity and conflicting expectations, particularly in rapidly changing educational systems. Balancing administrative tasks, teaching responsibilities and curriculum objectives can lead to stress, job dissatisfaction and difficulties in prioritising.

Assessment and accountability pressures

Increased emphasis on standardised testing and accountability measures can create psychosocial hazards for teachers. The pressure to achieve predetermined targets and meet performance indicators may lead to heightened stress, anxiety and a focus on 'teaching to the test' instead of holistic student development. The fear of negative consequences can adversely affect mental health and job satisfaction.

➡

OVER TO YOU

- What are three psychosocial hazards that impact your wellbeing at work?
- What strategies could you put in place to mitigate the impact of these hazards on your wellbeing?

Impact of psychosocial hazards on teachers

Exposure to psychosocial hazards in the workplace can have profound effects on our psychological health, job satisfaction and overall wellbeing.

Burnout

Excessive workload, emotional demands and a lack of support can contribute to a state of chronic exhaustion, cynicism and reduced professional efficacy. Burnout leads to increased absenteeism, decreased job performance and a higher likelihood of leaving the teaching profession.

Stress and mental health issues

Psychosocial hazards significantly contribute to stress among teachers. Long-term exposure to stress without adequate coping mechanisms and support may lead to the development of mental health issues, impacting both personal and professional lives.

Job dissatisfaction

When teachers face persistent psychosocial hazards, job dissatisfaction can arise. Reduced job satisfaction can lead to decreased motivation, a lower commitment to teaching and a decline in the quality of instruction.

Work-life imbalance

The demands of teaching combined with psychosocial hazards can create an imbalance between work and personal life. Teachers may struggle to find time for self-care, leisure activities and nurturing personal relationships, further exacerbating stress and impacting overall wellbeing.

Increased turnover and attrition

The negative effects of psychosocial hazards can contribute to higher turnover rates and attrition among teachers. When teachers experience significant job dissatisfaction and burnout, they may seek employment in other fields. This turnover negatively affects the continuity of education for students and the stability of the school.

Addressing psychosocial hazards

Mitigating psychosocial hazards promotes the psychological health of teachers. Let's look at some strategies that schools can implement to manage and prevent these hazards.

Workload management

Schools should strive to create realistic workloads that allow teachers to manage their responsibilities effectively. Reasonable class sizes and adequate time for lesson planning, grading and administrative tasks can alleviate stress and promote a healthier work environment. This means actively reviewing workplace systems that have become obsolete. Sometimes we have a tendency to do things how they have always been done, when in fact there may be better and more efficient ways of working.

Support and resources

Providing teachers with support opportunities such as mentoring programs, counselling services and quality professional development can enhance their growth and wellbeing. Adequate resources, materials and technology should be made available to ensure effective teaching practices.

Promoting a positive school culture

Schools should foster a positive and supportive culture that values and respects teachers. Encouraging collaboration, open communication and teamwork can create a sense of belonging, reducing feelings of isolation and promoting collegial relationships. Positive cultures do not happen by chance.

They require systems and processes that make people feel psychologically as well as physically safe. Observing how staff treat each other and work together provides a clear indication of a school culture.

Training on classroom management and emotional support

Training and professional development programs on effective classroom management, conflict resolution and emotional support can equip teachers with the skills to navigate challenging situations and maintain their wellbeing. When people feel good, they function well. When they have a sense of mastery over their experiences, they not only perform well but have a sense of agency to grow and learn.

Recognition and appreciation

It's vital to honour each other's efforts and accomplishments. Celebrating successes, acknowledging contributions and providing regular feedback can boost morale, job satisfaction and motivation. This is perhaps one of the easiest strategies for schools to focus on. It doesn't take much to authentically validate the hard work of others. This does not mean a generic global email for all to see, but a genuine interest in noting the time, effort and energy that goes in to managing everyday demands.

OVER TO YOU

- What workplace systems or platforms help you do your job?
- How do like to be acknowledged or recognised at work?
- How often are you given opportunities to engage in professional learning?

Policies and practices to reduce assessment and accountability pressures

Schools should implement policies that balance the emphasis on assessments and accountability with a holistic approach to education. Providing flexibility in curriculum implementation and supporting diversified student assessment can reduce undue pressure and its associated hazards. The design of work is most successful when we embrace as much flexibility as we can.

Just as consultation with all stakeholders is important, leadership support is a critical factor in any program's success (Durlak and DuPre, 2008). For teachers to deliver high-quality education, school policymakers have a responsibility to create the conditions in which they can function and perform well. Only with proper investment in time and resources will the psychological health of staff improve. This systematic process should lie at the foundation of any whole-school wellbeing program.

Real-life example

Carly began a new principalship at Sunny State School. She had recently attended my Inspired Leadership program, where the focus was on building connection before content. From the moment Carly set foot into her new school, she made it a point to get to know each and every teacher.

During one-on-one meetings with staff over the next few weeks, she took a genuine interest in each person's teaching methods, experiences and aspirations. Carly was committed to fostering a culture of appreciation and recognition where people felt valued for their hard work. She regularly visited classrooms, observing teaching practices and providing affirming feedback to teachers for the time and effort they put into their work.

It was a report-writing term, and Carly was aware of the related psychosocial hazards that could affect the wellbeing of teachers. As a result, she called a staff meeting to discuss options for reducing workload during this busy period. After a collaborative and open discussion, everyone agreed there would be no meetings for the three-week period. A report-writing workshop was offered and resources shared.

At the completion of this period, Carly proceeded to focus on staff recognition to showcase the teachers' accomplishments. Under her leadership, Sunny State School became a place where teachers felt valued, recognised and supported. The positive work environment fuelled their motivation, creativity and dedication to their students.

What does this mean for individuals?

Teachers can personally mitigate psychosocial hazards in the workplace by actively engaging in better work design practices. While some aspects of work design may be beyond individual control, there are several strategies that we can implement to reduce the impact of psychosocial hazards on our wellbeing.

Establishing clear boundaries between work and personal life is a good place to start. This means defining specific times for work-related activities, and prioritising self-care and leisure activities outside of work hours. Engaging in personal interests such as sport, art or crafts is a must.

It's also necessary to proactively manage our workload by prioritising tasks, setting realistic goals and breaking down complex projects into manageable steps. We can use time-management techniques such as to-do lists, calendars or planners and delegation whenever possible.

Actively seeking support from colleagues, administrators and other professionals within the educational community is important. It can be helpful to engage in collaborative problem-solving together as well as providing advice. A strong network is a source of emotional support that reduces isolation and creates a sense of belonging.

Foster open and effective communication with colleagues, administrators and parents. Clearly expressing concerns or challenges, sharing ideas and resources, and actively participating in discussions about work-related issues builds collective understanding. Effective communication enhances collaboration, reduces misunderstandings and creates a supportive work environment.

If you identify specific resource gaps or inadequate support systems that contribute to psychosocial hazards, you can advocate for change. By expressing your needs and concerns to administrators, you will play an active role in reshaping the work environment and securing necessary resources.

You should pursue opportunities for professional development to enhance your skills, knowledge and confidence in handling work-related challenges. Continuous learning and growth can increase job satisfaction and provide you with the tools to navigate psychosocial hazards effectively.

Conclusion

Psychosocial hazards in the workplace significantly impact psychological wellbeing, job satisfaction and retention in the teaching profession. By reflecting on these hazards, we can create a work environment that enables us to thrive. Recognising the importance of mental health and taking proactive steps to mitigate psychosocial hazards is not only beneficial for individual teachers but also for the quality of education provided to students.

The fertiliser to help us thrive

Teaching is an extraordinary profession. Teachers hold the power to shape young minds, ignite lifelong passions and create a lasting impact on society. We play a vital role in nurturing the potential within each student, encouraging growth and cultivating a love of learning. Yet challenges can leave even the most dedicated teachers feeling depleted and overwhelmed.

In Section 2, we delve into the secrets of thriving teachers. We uncover six concepts that act like fertiliser, providing the nourishment necessary for teachers to move beyond survival mode. These strategies serve as powerful tools to help us not only navigate the complexities of our profession but also find joy, fulfillment and professional growth along the way.

We will draw on the research discussed in the previous section to create a toolbox of practical applications. These activities give us an opportunity to merge theory ito action, so we can strengthen our internal and external resources to be well at work.

I chose the concept of thriving because it is known to be antidote to burnout (Spreitzer, 2021). THRIVE is an acronym that gives us a toolbox of resources to draw on. With these tools we will:

- Explore how we use **time**
- Observe how we think in our **head**
- Establish supportive **relationships**
- Notice our positive **impact**
- Connect to our **values**
- Regulate our **emotions**

Just as fertiliser enhances the growth and vitality of plants, the six concepts within the THRIVE model have the potential to revitalise our passion, resilience and wellbeing at work. Each concept represents a key element that, when implemented with intention and purpose, can significantly transform how we think, feel and respond.

But this book is not just a manual of techniques and practices. It is an invitation to embark on a transformative journey—one that encompasses personal growth, professional development and the realisation of you as a human being first and a teacher second.

While thriving is the aim, it doesn't mean perfection or constant happiness. It means having the awareness to meet your needs. It means being aware of your current demands and available resources. It means building your capacity to withstand and grow from challenges in your environment. It means revelling in and celebrating greatness when it occurs, and resting when we need to.

I invite you to play, be curious and have fun. Keep a journal of your insights and consider how the strategies can be integrated into your everyday routines. By implementing these strategies and nurturing them with intention and consistency, you have the power to not only survive the challenges of teaching but to thrive amidst them. Let us embark on this transformative journey together.

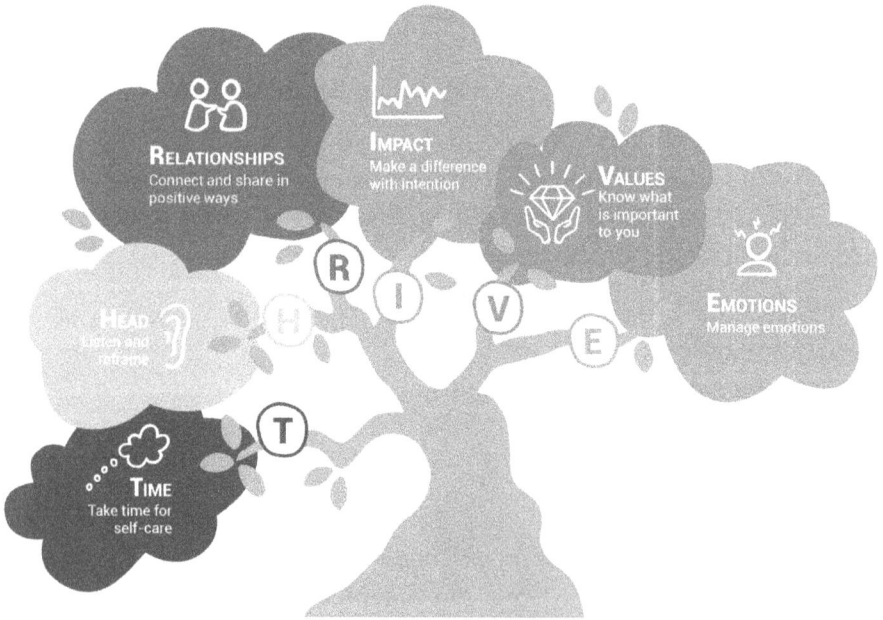

RELATIONSHIPS
Connect and share in positive ways

IMPACT
Make a difference with intention

VALUES
Know what is important to you

EMOTIONS
Manage emotions

HEAD
Listen and reframe

TIME
Take time for self-care

R H I V E T

Time

Thinking about time

Time is an ever-present factor in the life of a teacher. From managing classroom schedules and instructional planning to grading assignments and engaging in professional development, we constantly navigate the complex relationship between time and energy. How we think about time can significantly influence our effectiveness, wellbeing and overall teaching experience.

In this chapter, we delve into the intriguing realm of time perception and explore six strategies to help us think more effectively about this invaluable resource. The strategies will guide you through a process of introspection and self-awareness, encouraging you to critically examine your relationship with time and uncover underlying assumptions that may impact your experience at work.

Each strategy outlined in this chapter provides practical techniques to cultivate a mindful approach to time management. By incorporating these strategies into your daily routine, you can gain a renewed sense of control over your schedule, better prioritise your tasks and strike a balance between professional responsibilities and personal wellbeing.

By undertaking this journey of self-discovery and growth, we can revolutionise our understanding of time, unlock our full potential and forge a path towards more meaningful experiences at work. Lets embark on this temporal reflection and discover how to take back control of our time.

Stop spending it and start investing it

Time is fixed. We can't stop it, we can't store it and we can't get more of it. We have a finite amount of time each day, and we spend it. How we spend our time is our choice, even if it sometimes doesn't feel like it.

Have you heard the saying 'time is money'? Let's accept that. Take a moment to think about how you spend your money. Do you plan carefully and stick to a budget? Do you look at how much money you have and decide where it will go? Or do you not think about money and perhaps overspend it? Now let's think about time. Do you look at how much time you have and then allocate where you will spend it? Do you plan carefully and keep to your schedule? Or do you not think about time and find that it runs away from you?

I am pretty savvy when it comes to saving money, but I used to be more carefree about time. I would commit to helping by staying back late, talking on the phone to people who needed support and going out of my way to pick up sandwiches for a staff lunch. But it dawned on me that time is a currency in education. Time is precious, and we need to be more conscious about how we use it.

Before you buy something, you typically weigh up the pros and cons against your needs and benefits. You review your budget and make a conscious decision to go ahead with the purchase. But when it comes to our time, we feel like it is being taken from us. We tend not to give it the same attention as our financial decisions.

We count the time we use at the end of each day by reflecting on how busy we have been. We typically feel overwhelmed with so much to do that we feel busy just thinking about all the tasks on our list, even if we are sitting still and not actioning anything. Time ticks away while we sit motionless thinking about our deadlines. But when we have a sick child or a hole in the roof with rain falling in, we leap into action with clear priorities.

Instead of thinking about time as being spent or given away, let's look at it as being invested. What are you investing time in? Who are you investing time in? Is it worth the investment? How do you know? How much time are you investing in yourself?

When we say we don't have time for our wellbeing, we are really saying that it's not a priority. We do have time for what matters most to us. Unfortunately, we don't often put ourselves on this list. Instead we schedule our time by giving it away to everyone else. For example, the average Australian worker is 3.6 times more likely to prioritise work over their family (Duxbury and Higgins, 2008). It's also interesting to note that people claim to work more hours than they do because they are thinking about work (Duxbury and Higgins, 2008). Instead of waiting to fall ill or suffer extreme burnout, it's helpful to prioritise our wellbeing in micro-moments. If we are going to nourish our wellbeing, we need to invest time in ourselves.

So, what's the truth about time? Let's look at some numbers. There are 168 hours in a week (24 hours a day × 7 days a week). On average, teachers work 60 hours a week (40 contracted hours and 20 for free). We're asleep for about 50 hours a week (7 hours a night × 7 days). This leaves about 60 hours free.

You have time. The question is where are you spending these 60 hours and where would you like to invest them?

We need to shift how we think about time and start prioritising it for our own needs. What if you could prioritise one hour to nourish your wellbeing? This might involve 3 × 20min exercise routines or 6 × 10min meditations. It doesn't matter what you do, but you must invest in yourself.

It's wise to plan financially for the short and long term, and it's wise to do the same with our time. Just as we devise ways to pay for daily expenses, monthly bills and yearly holidays, we need to think about how we invest our time daily, monthly and annually. By doing so, we become less reactive to what is around us. We shift from feeling as if everything is urgent, and we become empowered by recognising we have choices. When we need money, we don't just hope it arrives randomly in our bank account; we find ways to make it. Time is the same. It won't just appear; you need to make time for what matters most to you.

Toolbox of activities

1. Permission slip

In a world of high accountability, we teachers know all too well the role of the permission slip. With this mind, why not start your journey of learning to thrive by giving yourself permission to prioritise your wellbeing?

→

OVER TO YOU

Consider this your permission slip to prioritise yourself. Transfer these words to your journal and fill in the blanks.

Permission to thrive

I, _____, give permission for myself to prioritise

my own wellbeing from _____ (date) as I learn to thrive.

Signed: _____ Date: _____

---

2. Time audit

Imagine for a moment that this book had no headings or punctuation, but consisted of words on a page with one concept rolling into the other. How would it feel to read? I don't know about you, but I would feel confused and overwhelmed. It would be hard to decipher where one concept started and another finished. We need headings and punctuation in order to process information.

Our day seems to roll from one activity to the next, with little punctuation to allow us to pause and transition. We run from task to task, rarely giving ourselves a moment to recover. We need to schedule time in our day to give ourselves the space we need to process our fast-paced world. We need to insert a comma signalling that we should pause and think. We need a full stop at the end of the day so we can rest, and we need close the chapter at the end of the week. Before we can do this, we must identify how we spend our time.

The good news is that teachers are very familiar with timetables and schedules. We allocate time by planning when to teach specific topics with scope and sequences. We schedule assessments, reporting, meetings and holidays. We colour-code calendars and set ourselves reminders. But what are we prioritising? Let's do an audit on how we spend our time before we start scheduling our priorities.

➡

OVER TO YOU

Create a table like the one below with time in one column, then a task in another. Over the next three days at work, record how you spend your time in 30-minute blocks. Identify the category each activity falls into. Feel free to make up your own categories.

Example

Time	Task	Category
6am	Up and get ready	Self
6.30am	Lunch prep	Family
7am	Drive	Travel
8am	Parent meeting	Work
9am	Teaching	Work

- What did you notice about where and how you spend your time?
- Which areas are you choosing to prioritise in your schedule?
- Where are the opportunities for you to schedule time for yourself?

Now that we've identified how you currently spend your time, we need to explore how you would like to spend your time.

OVER TO YOU

Let's begin by taking a blank sheet to schedule our time for the next three workdays, using a different pen colour for each action.

- First write down your non-flexible tasks (teaching timetable, meetings, duties).
- Next, list three to five important events for the week (family birthday, exercise, date night, creativity course).
- Then add travel, sleep and mealtimes.
- Now add any extras necessary for the week (lesson-planning time, admin time, research, networking).
- Finally, add in 'you' time. When and how will you rest and recover?

The goal here is to stop feeling like we are a victim of time and regain our power. If we don't consciously focus on what we want, we will inevitably end up with time passing us by, wishing our life away as we look forward to the next holiday.

Reflecting on what's important is a skill that can take some getting used to. One of the traps we fall into is thinking that everything is important and has to be done now. Sometimes this is driven by external demands, and sometimes it is caused by our own expectations.

People who don't know how to prioritise say they are unclear about what they want. If you feel uncertain about what areas to cut back on, list them all and reflect on your reasons for doing certain tasks and your intentions for their impact. This may help identify whether these tasks are priorities or false truths. It's time we scheduled our priorities instead of prioritising our schedule.

3. Daily ME-ting for a self-care check-in

A few years ago, I heard an excellent tip from British life coach Jay Shetty about meetings. He was reflecting on the endless meetings he had each day and how much time they took away from things that made him feel alive. Playing around with words, he realised he needed a 'ME-ting'. He decided to schedule into his calendar a 30-minute ME-ting each day to recharge his battery. This could be done by sitting, walking, socialising or breathing deeply or whatever else he felt like doing. It was his priority to show himself that he cared about his own worth.

I mentioned this to a colleague who was a professor of education at a large tertiary institution. She struggled with finding time for herself because all staff calendars were global, meaning that people knew where everyone was at all times and could openly book meetings if a time was vacant. She felt she couldn't insert ME-ting into the calendar, so she wrote 'Jim' instead. This became her twice-weekly gym session that she knew she needed. It was a ME-ting that became a non-negotiable on her calendar.

→

OVER TO YOU

Now it's time for you to plan time for your own ME-ting. It can be short or long. To help you, I've created a Triple-A action plan for self-care that requires you to stop and ask yourself three simple questions.

1. Assess

What is happening for me right now? *I feel tension in my shoulders with cramps in my neck.*

2. Awareness

What do I need? *I need to stand up and stretch.*

3. Action

When will I do this? *I'm in a meeting now, but in 20 minutes I will stand up and stretch.*

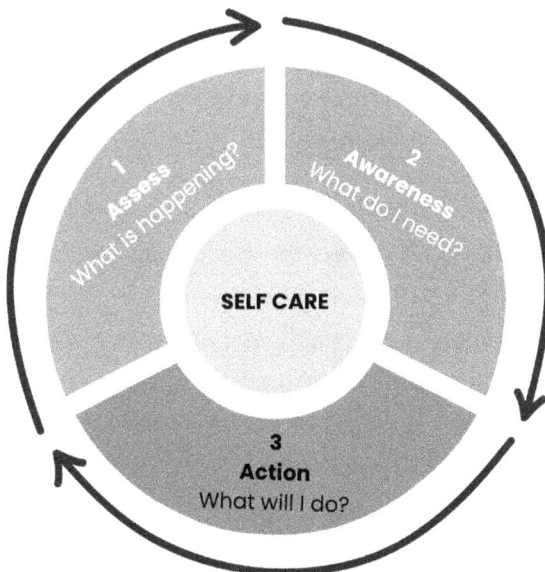

Keep the Triple-A action plan near you and check in with yourself often. It's quick, simple and a good way to action self-care.

4. Use a list-making system

How do you currently organise the tasks that need to be completed today and in a week? How do you manage this list in order of importance, and how do you track what is completed?

I am still shocked when I hear teachers say they keep a lot of information in their heads. No wonder they don't sleep well. Sometimes I see people with to-do lists that are two pages long and feel overwhelmed for them, knowing there is no way they can complete all the tasks in time. We need to be better organised to free up thinking space. We need to be realistic about what is possible to achieve in a day and set ourselves up to enjoy our job.

US teacher Angela Watson has created an excellent list-making system as part of her 40-Hour Teacher Workweek. Her weekly to-do list is broken up into seven days. Each day has a section to write down what you do before school, at midday, after school and at home. At the bottom of each day is a section to record things you need to follow up on or find out about. By breaking up our list into these sections, we set more realistically achievable goals. If you know you have a full teaching day ahead, the likelihood of achieving any planning or admin is going to be quite low. Prepare yourself for success, not overwhelm.

I like to make a list on Sunday evenings of everything that is happening in the week ahead. On the relevant pages in my diary, I write down fixed events or meetings. Then, based on time available, I write in other tasks with a goal I would like to achieve: writing 300 words for a new blog, preparing for a podcast interview, following up with a principal. As the week unfolds and new tasks come to light, I determine their level of urgency and add them to a day of the week when I know I have the space to deal with them. You can't do everything at once; some things just have to wait.

It doesn't matter which system you use, as long as you choose something. Stop waking up in the middle of the night with your head racing. Stop trying to rely on your memory. Your mental capacity can't cope. And stop filling up your to-do list with things that will take you all term to complete.

OVER TO YOU

Take a moment to think about the systems you use to meet workplace demands. These could include gradebook software to track student progress, Learning Management Systems (LMS) and classroom-management apps to aid in behavioural tracking and engagement. You probably also employ to-do lists, calendars and digital note-taking tools. Collectively, these systems streamline the complex demands of teaching to enhance productivity and learning.

- Which systems do you use to organise your day?
- Which ones are working well for you?
- Are there any that need reviewing or improving?

5. Shut-down ritual

Computer science researcher Cal Newport talks about the science of productivity in his book *Deep Work* (2016). He explains how our increasingly distracted world hinders our ability to both concentrate and switch off. Newport believes that the problem lies not in the distractions themselves but rather in our inability to manage them. Instead of prioritising what's important to us, we become controlled by our response to distractions. This steals our time and attention.

Constant interruptions, connectivity to devices, open-plan offices and expectations of fast responses leave little time for cognitive rest. We move from one cognitive task to the next with little regard for downtime. We need to create a shutdown ritual, a signal to our brain that it is time to switch off from work and switch on to rest. This means no after-dinner email checking, no mental replays of conversations and no scheming about how you'll handle an upcoming challenge. As best as you can, close off work thinking completely. We need to rest to recharge the energy needed for the next day's high cognitive load.

Your brain is engaged in tasks, decisions and problem-solving throughout the workday. This leads to cognitive overload, which makes your mind feel cluttered and overwhelmed. A shutdown ritual allows you to intentionally unload your cognitive burden by wrapping up tasks and mentally preparing to transition into personal time. This can alleviate stress and improve mental clarity.

Productivity isn't about putting in more hours. It's about getting the most out of the hours you work and the most recovery out of the hours you don't. Implementing a shutdown ritual establishes a clear boundary between work and personal life. It helps prevent work-related thoughts and stress from spilling over into your personal time, allowing you to fully engage in non-work activities. This separation contributes to a healthier work-life balance, reducing the risk of burnout and promoting wellbeing.

For example, I will always pack my bag the night before any event. As I drive to the school, I listen to music that I find energising (usually anything from the 80s). When driving home, I sit in silence to reflect on the day and restore my mental energy. At home, I go straight to my office, unpack my bag and check my emails for anything urgent. Most things I write in my diary to action the next day or the following week. This is not 'work time', but 'wrap-up time' that clears my mind and prepares me for the next day. I don't hold anything in my mind because I have written down my plans for the day ahead. I then close my diary, turn off my computer and shut my office door. Finally, I change my clothes, return to my family and settle in for the night of banter, TV or whatever else might be happening. I don't check emails or social media and I don't accept phone calls. The evenings are my time to rest my mind on stuff that interests me. I'm not saying you should do this, because everyone's context and needs differ. But having a start-and-finish ritual to your day can be a great way of signalling to your brain that it's time to switch on or off.

OVER TO YOU

Think about an average workday and reflect on the following questions:

- What rituals do you use to signify the start your day?
- How do you currently switch off at the end of a workday?
- If you don't have any current rituals, what could you do?

6. Healthy boundaries

So much of our life feels like we are responding to the things happening to us. It's easy to forget that we have choices. We choose to spend our Saturdays running our children around to sporting activities. We choose to take on leadership roles. We choose to go or not go to the gym.

The problem is that teachers tend to say 'yes' too much and 'no' too little. We see people in need and enjoy being of service. Most of us are people-pleasers who go above and beyond. We tend to over-commit and get frustrated when we can't meet our own expectations. Every time we say 'yes' to one thing, we are saying 'no' to another. This usually means ourselves.

Creating healthy boundaries begins with saying 'no'. You can't do everything for everyone all the time. You're not a machine, remember? Being clear on our boundaries is essential if we are going to give ourselves more time to do things that help us thrive. This doesn't mean sending out ultimatums to people or drawing lines in the sand. It means communicating in ways that are clear, open, honest and respectful. For example, imagine that a parent has a habit of emailing you on weekends. You feel pressured to respond as you know they expect this from you. Instead, you politely email them back on Monday, letting them know that they are welcome to contact you anytime but to expect your reply during office hours.

Another example could be a colleague who frequently interrupts you during work that requires deep concentration. Instead of huffing and puffing in frustration, you could politely let them know you need the next 50 minutes to concentrate and would be happy to go to another room if that was preferable. This may seem harsh at first, but keep in mind that schools are workplaces where it is reasonable for people to set boundaries around completing necessary work as part of their professional conduct.

Now it's your turn. Have a look at these five steps of constructive boundary-setting:

1. **Express appreciation and gratitude.** Start by expressing gratitude for being considered and acknowledged for the additional task. For example, say: 'Thank you for thinking of me for this opportunity.'
2. **Clarify your current commitments.** Politely explain the existing workload or responsibilities that are occupying your time. You can say:

'At the moment, I have several ongoing projects and commitments that require my full attention.'

3. **Highlight your priorities.** Emphasise the importance of your current responsibilities and the impact they have on your students or the educational program. For instance, say: 'I want to ensure that I'm giving my best to the current projects I'm involved in, as they directly contribute to student learning.'

4. **Offer an alternative solution.** Instead of outright refusing, propose a solution that may help meet the request or suggest someone else who could be a suitable fit. For example, say: 'I'm unable to take on this task at the moment, but I can recommend a colleague who might be available and well-suited to it.'

5. **Express willingness to help in the future.** Politely communicate that while unable to take on the additional work now, you are open to future opportunities or to be considered for tasks that align better with your current workload. For instance, say: 'I appreciate the offer, and I would be happy to contribute in a different capacity when my schedule allows.'

→

OVER TO YOU

Now that we have explored ways to set clear and respectful boundaries at work, think about which method will be most useful to you. What could you choose and why?

7. Tiny habits

Habits are our brain's way of reducing cognitive load. By initiating a behaviour that doesn't require a conscious decision, we are saving brain power. Given that cognitive fatigue is a major contributor to our feelings of overwhelm, it is helpful to reflect on our habits and determine if these are helpful or harmful. The next step is to build healthy habits.

Social scientist BJ Fogg has researched why and how people change behaviour, introducing the concept of 'tiny habits' in his 2019 book of the same name. The concept revolves around the idea that small, incremental behavioural changes are more effective in creating lasting habits than big, drastic changes. The Fogg behaviour model suggests that three elements need to converge simultaneously for a behaviour to occur: motivation, ability and a prompt. Motivation can fluctuate, but ability is a critical factor that can be improved to make behavioural change easier (Fogg, 2019). It's important to note that habits are not goals. Goals are outcomes and habits are how you get there.

Start by identifying the behaviours you want to cultivate or change. Be specific about what you want to achieve. Design tiny habits that require minimal effort and are easy to integrate into your daily routine. These habits should be so small that they're almost laughably simple. Instead of committing to an hour-long workout, start with doing one push-up or taking a five-minute walk.

Anchor your tiny habits to an existing routine or event in your day. This helps establish a trigger or prompt for the behaviour. If you want to develop a habit of drinking more water, anchor it to eating. When you feel hungry, commit to taking just one sip.

Celebrate your successful completion of the tiny habit by acknowledging your accomplishment, even if it seems insignificant. Celebrating reinforces the positive feeling associated with the behaviour, making it more likely to be repeated in the future.

Once the tiny habit becomes automatic and effortless, you can increase the difficulty or duration. If your initial tiny habit was one push-up, you can increase it to two or three push-ups. By focusing on small, achievable changes, the concept of tiny habits encourages a positive and sustainable approach to behavioural change.

Put simply, a tiny habit is a simple recipe with only three ingredients:

1. **An anchor.** A moment or a habit that already exists in your life. Getting out of bed, brushing your teeth, having breakfast.
2. **A tiny behaviour.** A very small version of a larger behaviour you want to create. A larger behaviour may be to floss all your teeth. The tiny version would be to floss one tooth.
3. **A celebration.** The emotion that will help make your habit memorable and stick.

OVER TO YOU

Habit change requires conscious thought and action. Take a moment to think of a small incremental change you would like to make at work. It could be taking more deep breaths, eating healthier snacks or using the Triple-A action plan of self-care.

Take the three ingredients of a tiny habit to plan how you could build this into your day. For instance: *When the bell goes at the end of each period, I will place my hand and my chest and take three deep breaths. I will celebrate this action with a smile on my face!*

Head

Looking inside

Our inner world is filled with thoughts, beliefs and perceptions that shape our experiences and influence our actions. At the centre of this landscape lies self-talk, the internal dialogue that accompanies us throughout our lives. It is through self-talk that we interpret events, form judgments and construct our sense of self. However, not all self-talk is constructive or supportive. Often our inner critic emerges to undermine our confidence, fuel self-doubt and impede our personal growth.

In this chapter, we explore six strategies that can help us better manage our inner critic. By cultivating an awareness of our thoughts and developing strategies to navigate the critical voice within us, we can transform our relationship with ourselves and unlock our potential.

The strategies presented here are designed to guide you through a journey of self-discovery and introspection that empowers you to notice, challenge and reframe negative self-talk. By understanding the roots of our inner critic and its impact on our wellbeing, we can develop the skills to quiet its negative influence and replace it with self-compassion, self-acceptance and self-empowerment.

By exploring cognitive reframing and other evidence-based approaches, we can gain a deeper understanding of our thought patterns and the ways in which they shape our perceptions of ourselves and the world around us.

Self-care will enable us to cultivate a sense of resilience, enhance our self-esteem and foster a positive mindset. In this nurturing space the inner critic loses its power, allowing us to embrace our authentic selves and pursue our goals and aspirations with confidence and determination.

The strategies presented here offer a pathway to liberation from the grip of self-doubt, negativity and perfectionism. Let us embark on an enlightening discovery of self-awareness .

Don't believe everything you think

Did you know we have 60,000 thoughts a day? For many of us, a significant number of these thoughts come from an inner critic. This internal dialogue can be harsh and damaging.

I have many voices in my head. One is that of hope and possibility, but then the voice of self-doubt pops up to remind me of what could go wrong. There have been many days where I oscillate between the two in a tussle for control. What I have learned is that if my inner critic is left unchecked, I rob myself of enriching opportunities to learn and grow.

When I became the head of a faculty, I found myself very nervous about making poor decisions or not knowing how to answer the questions people would ask me. I questioned myself often. My inner critic was frequently telling me to do more, be more and work harder. Thankfully, my years of therapy and self-discovery helped me tame these thoughts and allowed me to be a better friend to myself.

We all have automatic negative thoughts, also known as ANTs. These thoughts stem from our beliefs and past experiences. They often come in the form of 'I'm not good enough', 'I'm not smart enough' or 'I'm not doing enough'. These thoughts push you to your limits and deteriorate your self-worth. It is important to understand that they are not facts. Don't believe everything you think!

Our perspective is driven by the meaning we've made from our past experiences. For example, I have always loved dogs. When I see a dog, I immediately equate it with a feeling of playfulness and joy. My sister, on the other hand, had a negative experience with a dog when she was younger. In her mind they equal alertness and danger. We may see the same dog at the same time, but our responses will not be the same. I will move toward the dog, and she will move away from it. What we think influences how we feel, which then influences how we choose to respond. It can be helpful to question our thinking as a powerful way to support behavioural change. This is called cognitive reframing.

Cognitive reframing

We make assumptions about events to give them meaning. Called schemas, these thought patterns help us understand how the world works and where we fit in it. Schemas are not necessarily truths, but are based on our perception of reality. These thought patterns influence our feelings, which in turn influence our behaviour. In cognitive behavioural therapy this is known as the thought-feeling-behaviour loop. Emotions and behaviour are not triggered by events themselves, but by how we interpret these events.

There may be several teachers in a staffroom, all hearing and seeing the same thing. Their perceptions will differ, and so will their responses. Imagine that two people have been asked to take on extra playground duty for someone who is absent that day. One person thinks, 'Why should I? I don't have time for this.' These thoughts stay with them for the rest of the day. At playground duty they feel frustrated and resentful. As a result, they may be harsh or reactionary with their students. The other person thinks, 'I really don't have time for this right now, but it gives me an opportunity to catch up with a few students I've been wanting to talk to.' They acknowledge that the situation is not ideal and reframe how they look at it. Instead of holding onto the frustration, they are able to acknowledge and shift their perspective to one of acceptance and possibility.

Cognitive reframing is frequently used by therapists to restructure unhelpful thinking. It aims to help people reduce their stress by cultivating more positive and functional thought habits (Mills, Reiss and Dombeck, 2008). When we change our thoughts about something, we change our experience of it. This is not about ignoring negative emotions, but rather about questioning our perspective. That's especially important for those of us who ruminate over past events by playing them over and over in our minds. Replaying experiences without solutions tends to make our reaction disproportionate to the actual event. In this way, rumination can prevent happiness and negatively impact our wellbeing (Nolen-Hoeskema, 1993).

The first step in cognitive reframing is to notice your thoughts. This can be tough when you are busy, but it's an important skill that is worth taking the time to practice. Acceptance and commitment therapist Russ Harris uses a great metaphor to explain this (2017). Imagine that you're at a sushi train restaurant where plates of sushi go past you on a conveyer belt all day long. Imagine that the sushi on those belts are your thoughts, and the chef who makes the sushi is your mind. Some of the sushi is very appealing and some is not. Some thoughts are very pleasant, perhaps making us feel energised. Others can be hurtful or harmful. Just as in a sushi train, we don't have to consume everything that comes past us. We can choose which thoughts to hold onto and which ones to let pass by.

Once we notice our thoughts, we need to identify whether they are helpful or harmful. Thoughts are messages to help us build meaning. By asking ourselves whether a thought is helpful or harmful, we get to choose the voice of our inner critic or our inner friend. If we have identified a thought as harmful, we then have the opportunity to challenge it. We can do this by asking ourselves if there is another way to look at the situation.

Take a moment to explore the following activities that will help you hear your inner voice and better manage it.

Toolbox of activities

1. ANTs and PETs

In cognitive behavioural therapy, the concepts of ANTs (automatic negative thoughts) and PETs (performance-enhancing thoughts) are used to explore and modify thought patterns that influence emotions and behaviours. As we've learned, ANTs are the often-distorted thoughts that automatically arise in our minds in response to certain triggers. These thoughts are typically irrational, self-critical and pessimistic, contributing to negative emotions and maladaptive behaviours. ANTs include beliefs such as 'I'm a failure', 'I'll never succeed' and 'nobody likes me'.

PETs are the opposite of ANTs. PETs are positive, affirming and rational thoughts that we can cultivate to counteract negative thinking patterns. PETs include thoughts like 'I can handle this challenge', 'I've overcome obstacles before' and 'I am capable and deserving of success'.

The process of identifying and challenging ANTs while nurturing PETs is a key component of cognitive behavioural therapy. Through cognitive reframing techniques, we can learn to recognise negative thoughts, evaluate their validity and replace them with more realistic and empowering thoughts. This helps us gain control over our thought processes, improve emotional regulation and develop more effective coping strategies.

This is more than positive thinking. It's about questioning our thoughts to check if they are helping us negotiate challenges as best we can. This begins by catching our ANTs, stomping them out and replacing them with PETs.

OVER TO YOU

Here are some ANTs that I typically hear in schools:

- *I'll never be able to get through to this student.*
- *It's not fair that I have to do an extra playground duty this week.*
- *We've already tried this, it didn't do anything.*
- *That will never change around here.*

How could these ANTs be reframed?

For example:

- *I'll never be able to get through to this student... Or maybe the student is simply doing the best they can at the moment given their context at home.*

Take a moment to think about what's happening in your head by reflecting on some common ANTs you tell yourself at work. Write them down in your journal.

Looking at your list, take a moment to reframe each statement from into a PET.

2. Shut down imposter syndrome

Have you ever felt like a fraud because you secretly didn't know what you were doing? There were many times when I felt like a fraud. Despite other teachers telling me how effective I was at managing difficult students, I always felt like I was winging it. I didn't really know what I was doing and then felt uncomfortable receiving praise for it. I later came to learn that this was called imposter syndrome.

Imposter syndrome is a psychological phenomenon by which we doubt our own accomplishments and fear being exposed as frauds, despite evidence of our competence. Those experiencing imposter syndrome often attribute their success to luck or external factors, discounting their abilities and feeling like they don't deserve their achievements. This self-doubt can lead to anxiety, fear of failure and a constant need to prove oneself. Imposter syndrome commonly affects high achievers, including both students and teachers. This way of thinking can be debilitating if it goes unchecked.

Imposter syndrome can undermine self-efficacy, as we come to question our abilities and struggle to internalise our accomplishments. Developing self-efficacy involves recognising and overcoming imposter syndrome, fostering a belief in our capabilities and embracing a positive self-perception.

Have you ever had the following thoughts?

> *I have no idea what I'm doing.*
>
> *I'm just not capable of doing everything that needs to get done.*
>
> *I don't have the experience to do this job well.*
>
> *Other people seem to be managing so much better than I am.*

Other people aren't lying to you when they identify the great work you're doing. The issue is that you don't think you're as good as you are. This usually occurs when your inner critic is loud or unchecked.

Imagine for a moment that your best friend has come to you because they are feeling down. They are being unnecessarily hard on themselves despite doing a great job. I imagine you would challenge their thinking and remind them of all their recent achievements, as well as the skills and talents they possess. You would be kind, compassionate and supportive. This is the job of our inner friend too. Next time you catch yourself with imposter syndrome, imagine you are your own best friend. Turn off the radio station of the inner critic and turn up the volume of being your own best friend. How would your best friend talk to you? What would they say to shift your perception of yourself?

→

OVER TO YOU

Take a moment to reflect on a time when you may have suffered from imposter syndrome by answering the following questions.

- Make a list of five statements that your inner critic says to you about work. When you read these out loud, how do they make you feel?

- Are these statements helpful or harmful to your overall wellbeing?

- Now make a list of five statements that your best friend would say to show you compassion, kindness and care. When you read these out loud, how do they make you feel?

- Are these statements helpful or harmful to your overall wellbeing?

3. Reframe learned helplessness

Learned helplessness is a theory of depression introduced by positive psychologist Martin Seligman in 1972. It describes a psychological state in which individuals feel powerless after being conditioned to expect pain, suffering or discomfort without a way to escape (Cherry, 2017). This pain can be physical or emotional. It can involve disappointment, frustration, blame, shame and feelings of guilt. If we experience enough conditioning of these feelings, we may come to expect them as normal and stop looking for ways to remove or limit the pain. When we come to understand (or believe) that we have no control over what happens to us, we begin to think, feel and act as if we are helpless.

The key word here is 'learned'. No-one is born believing that they have no control over what happens to them, that it is fruitless to even try to gain control. This is a learned behaviour that can equally be unlearned. A teacher who continually faces challenging classroom behaviours but feels powerless to address them may experience learned helplessness. Despite attempting various strategies and interventions, this teacher will perceive their efforts as futile and resign themselves to the belief that they have no control over student behaviour. This in turn hinders their ability to proactively find solutions.

A teacher who consistently receives negative feedback from a leader may also experience learned helplessness. Over time, this teacher internalises the criticism and starts to believe themselves inherently incapable of meeting expectations or making positive changes. This learned helplessness leads to a lack of motivation, reduced engagement and a diminished belief in their ability to grow and succeed. It's important that we recognise and address these feelings by seeking support, exploring alternative strategies and cultivating a growth mindset to regain a sense of agency and empowerment in our professional lives.

OVER TO YOU

Self-awareness is foundational to social and emotional competence. Reflect on your thinking by answering the following questions:

- Have you ever thought to yourself: 'There's no point in doing this because it won't work anyway'?
  - What circumstances led to this?
  - What were you thinking at the time?
  - How did this make you feel?
- Have you ever thought to yourself: 'I'll give it a go and see what happens'?
  - What circumstances led to this?
  - What were you thinking at the time?
  - How did this make you feel?

4. Notice your thinking traps

Also known as cognitive biases, thinking traps are patterns of distorted or irrational thinking that lead to errors in judgment and decision-making. These distortions are mental shortcuts used by our brains to simplify complex information processing. They can result in flawed reasoning. Thinking traps limit our ability to consider alternative perspectives, impede problem-solving and dampen critical thinking. We must recognise and challenge thinking traps to enable optimal decision-making and a more accurate understanding of the world.

Have you ever thought someone doesn't like you without ever finding out if it's true? If you have, you may have been suffering from one of the many thinking traps or cognitive distortions that can hijack your brain.

Common thinking traps include:

- **Fortune-telling** – predicting the future with little evidence for our conclusions.

  I'll never be able to get through to this student no matter what I do.

- **All-or-nothing thinking** – otherwise referred to as black-and-white thinking.

  If I can't do it right, what's the point in trying?

- **Over-generalising** – seeing a pattern based on one event or being overly broad in our conclusions.

  Using phrases such as 'everybody thinks…', 'I never…' or 'I always…'

- **Emotional reasoning** – assuming that because we feel a certain way it must be true.

  I feel lonely, which means I am unsupported by others.

- **Mind-reading** – imagining what others must be thinking or feeling.

  My colleague thinks I am lazy.

- **Catastrophising** – blowing things out of proportion.

  There is no way I can get up at assembly and speak, I'll totally freak out!

- **Should statements** – telling yourself you 'should' know or do something, resulting in disappointment

  I should have known better than to try that new activity.

➔

OVER TO YOU

- Choose a recent situation in your teaching experience that was particularly challenging. It might be an interaction with a student, a parent-teacher meeting or a decision-making process.

- Recollect your initial thoughts and reactions during the situation. What were the immediate thoughts that crossed your mind? What emotions did you experience?

- Which of the aforementioned thinking traps did you fall into?
- Reflect on how these thinking traps influenced your thoughts, emotions and behaviours. Consider how they may have hindered your ability to see alternative perspectives, problem-solve effectively or maintain a balanced approach.
- Take each identified thinking trap and challenge it with more rational and balanced thoughts. Actively seek out evidence or viewpoints that contradict your initial assumptions.
- Identify specific actions or strategies you can implement to minimise the influence of thinking traps in your teaching practice. Commit to practicing critical thinking, seeking diverse perspectives and regularly examining your thoughts for biases.

5. Challenge your thoughts

Once we are able to notice a thought and reflect on whether it is helpful or harmful, we can then challenge it to shift our perspective and emotional response. In 1955, psychologist Albert Ellis created a model at the root of cognitive behavioural therapy (Ellis, 1957). The ABC model is a cognitive framework that helps us understand how our beliefs and interpretations of events influence our emotional and behavioural reactions.

The model explains that an external event (A—activating event) does not cause an emotional response, but the beliefs (B) associated with this event do. It is these beliefs that result in our conclusions and actions (C—consequences). When we question our beliefs, we can shift consequences and change how we think and feel.

Martin Seligman built on this to create the ABCDE model of optimism (2006). This too is a cognitive restructuring tool used in cognitive behavioural therapy. When we experience challenges or adversity (A), we draw on existing beliefs (B) for an explanation. As we build meaning around this, emotions appear and actions occur (C). To overcome unhelpful thinking, we can dispute (D) the belief to bring about a more effective, energising or empowering (E) outcome.

The ABCDE model helps us recognise and challenge negative thoughts that contribute to distressing emotions. By reframing and replacing these thoughts with more balanced and realistic alternatives, we can gain better control over our emotions (Beck, 2011). This process promotes critical thinking and fosters effective problem-solving skills, enabling us to approach challenges and difficulties in a more adaptive and proactive manner (Beck, 2011).

These models facilitate self-reflection and introspection, allowing us to become more aware of our automatic thoughts, cognitive biases and underlying assumptions. This heightened self-awareness can lead to a deeper understanding of our thinking patterns and their impact on wellbeing, fostering personal growth and self-discovery (Dryden and Neenan, 2004). By challenging and reframing negative thoughts, we build our resilience and ability to cope with adversity. Viewing setbacks and failures as learning opportunities promotes a growth mindset and fosters psychological resilience (Yeager and Dweck, 2012).

Let's look at an example of how we can turn around our thinking using the ABCDE model.

- **A**dversity. You conducted a professional learning workshop with your team at school. You were lost for words in a few places and you're not sure if people found it helpful.

- **B**elief. *I stuffed up. I'm hopeless at presenting to colleagues. I'm not doing that again.*

- **C**onsequence. You don't follow up by asking people questions a few days later. You don't offer to do any further presentations at work.

- **D**ispute. *I probably tried to get through too much, and I haven't really had a lot of practice running professional learning sessions with colleagues. People did seem interested and asked some questions. I might not have been completely articulate, but I did manage to get most of the key messages across.*

- **E**mpowerment. *Next time, I'll cut out some content and give myself more time to engage people with activities and questions so I can get instant feedback as I'm presenting.*

OVER TO YOU

Consider a challenging situation at work. It could be a difficult conversation with a parent, frustration at a student's lack of effort or overwhelm caused by a looming report-writing deadline. Now use the ABCDE model to change your thinking.

Adversity

- What happened in the situation?
- Who was involved?

Belief

- What were you thinking at the time?

Consequence

- How did these thoughts make you feel?

Dispute

- Are these thoughts realistic? How do you know?
- What could be some more helpful thoughts?

Empowerment

- How do you feel when you consider these helpful thoughts?

6. Embrace a growth mindset

A growth mindset refers to the belief that abilities, intelligence and talents can be developed through effort, learning and perseverance (Dweck, 2006). Teachers and students with a growth mindset embrace challenges as opportunities, view setbacks as learning experiences and believe that with dedication and practice they can improve their skills and abilities. This mindset fosters a love of learning, resilience in the face of obstacles and a willingness to take on challenges.

When facing difficult situations such as managing classroom behaviour or implementing new teaching strategies, we are more likely to persist and seek solutions with a growth mindset (Yeager and Dweck, 2012). This builds resilience, engenders a sense of accomplishment and reduces feelings of helplessness.

A growth mindset can motivate us to actively pursue professional development to acquire new knowledge, skills and pedagogical approaches (Hong et al.,

1999). Engaging in continuous learning and seeking opportunities for growth supports self-efficacy and job satisfaction.

Finally, a growth mindset can help us manage stress and cope with setbacks. We come to view challenging experiences as valuable sources of information for improvement rather than personal failures (Grant and Dweck, 2003). This perspective reduces stress and allows us to maintain a positive outlook, supporting our wellbeing in demanding environments.

OVER TO YOU

- Think of a teaching situation where things did not go as planned. It could be a challenging classroom situation, a lesson that didn't engage students as expected or any other relevant experience.

- Take a moment to remember the factors that contributed, and any emotions or thoughts you experienced during and after the incident.

- Reflect on the concept of a growth mindset and our potential for learning and improvement. Consider the insights you gained from the experience. What did you learn about yourself as a teacher?

- What strategies or adjustments could you employ in the future to prevent similar mistakes or handle challenges more effectively?

- Based on your reflections, outline specific actions or strategies you can implement. How can you apply what you have learned to improve your teaching practice?

Relationships

The power of connection

Healthy relationships at school significantly affect a teacher's wellbeing. The quality of the connections we cultivate with students, colleagues and parents greatly influence our job satisfaction, resilience and mental health. Just as we need food, water and air, we also need social relationships to thrive (Diener and Biswas-Diener, 2008).

Humans are wired to connect. Healthy relationships allow us to grow and learn, heal from setbacks and contribute to a common purpose. Having someone to share ideas with, challenge our perspective and value our contribution builds meaning. Positive connections develop core internal resources including social and emotional skills (Roffey, 2012). The better the connection, the greater our sense of belonging to something greater than ourselves.

The emotions that fuel relationships influence our productivity, performance, engagement and wellbeing. We know that people are drawn to those who are happier and more optimistic at work (Carver, 2010). A Gallup survey suggested that people who had a best friend at work were 43 per cent more likely to report having received praise, and 37 per cent more likely to feel supported and encouraged at work (2008).

In schools, where time moves fast and information needs to be shared quickly and widely, messages can become skewed. Communication plays a key role in the formation and maintenance of school relationships.

How we send and receive information can be open to interpretation, with the potential to leave people feeling valued and supported or judged and patronised. Self-awareness and self-management are crucial if we are to form positive relationships. Teachers need to have a myriad of skills to navigate personality types, communication styles and emotional triggers. We must be able to listen effectively, show empathy, read cues and adjust our approach based on context.

We will now delve into the importance of building healthy relationships at work, with six practical strategies that can empower you to foster a supportive and uplifting professional community. These strategies extend beyond mere camaraderie into meaningful connections that promote collaboration, empathy and mutual support. They will enable you to strengthen your wellbeing while contributing to a thriving educational environment. Together we can create a community that not only lifts us up personally but also fosters a rich and fulfilling educational experience for everybody.

Personality clashes

Personality clashes are common in any workplace. They can occur due to differences in communication styles, work preferences, values and approaches to conflict resolution. In my experience, they fit into three different categories:

1. The perfectionist vs. the easygoing

A teacher who strives for perfection and strict adherence to rules may clash with a colleague who has a more relaxed and easygoing approach to work. This clash can arise from differences in expectations, time management and approaches to classroom management.

2. The introvert vs. the extrovert

A teacher who prefers solitude and quiet may clash with a colleague who thrives on social interactions and group activities. This clash can manifest through differences in communication styles, collaboration preferences and even classroom dynamics.

3. The innovative thinker vs. the traditionalist

A teacher who is open to innovative approaches may clash with a colleague who prefers to stick to traditional teaching methods. This clash can arise from differences in pedagogical beliefs, adaptability and eagerness to embrace new technologies or teaching strategies.

Personality clashes can create tense and stressful work environments. Conflict and tension overshadow the positive aspects of our work, making us less engaged and fulfilled. When we do not get along well, it becomes challenging to share ideas and provide mutual support. This lack of collaboration can limit professional growth opportunities and reduce the sense of belonging within a school community.

When conflicts remain unresolved or escalate, they negatively affect team morale and cohesion. Prolonged exposure can contribute to job strain and burnout. Dealing with constant tension takes a toll on our emotional energy and resilience, making us more susceptible to burnout.

It is crucial we address personality clashes to build a culture of empathy and understanding among colleagues. This requires a willingness to understand as well as be understood. Social and emotional skills can support this.

Toolbox of activities

1. Show you care

Empathy is our ability to understand and share the feelings, perspectives and experiences of others. Empathy fosters a supportive and compassionate environment by promoting positive relationships, reducing stress and enhancing job satisfaction.

Research suggests that empathy plays a crucial role in teacher wellbeing. A study by Jennings and Greenberg (2009) found that when teachers perceive higher levels of empathy from their colleagues and administrators, they experience greater job satisfaction and lower levels of stress. Empathetic interactions create emotional connection and support, which contribute to a sense of belonging and psychological wellbeing (Eisenberg et al., 2019).

Furthermore, a study by Reinke et al. (2011) demonstrated that teacher empathy positively influences student-teacher relationships and the classroom climate. Empathetic teachers create supportive learning environments that make students feel valued and understood (Pianta, 2018). Empathy fosters a sense of safety, trust and respect, enhancing both student and teacher wellbeing.

Empathy also sets the foundation for a positive organisational culture. When embraced as a core value, it shapes workplace culture to prioritise understanding, respect and support. When colleagues and supervisors demonstrate empathy, it creates a sense of emotional connection and support. Feeling heard and valued boosts our satisfaction. When we are understood, we are more likely to communicate openly, resolve conflicts constructively and work together to achieve shared goals. Positive interactions based on empathy enhance teamwork, reduce stress and foster a sense of belonging.

OVER TO YOU

You may have heard the saying 'people don't care how much you know until they know how much you care.' How do you show people you care?

Here are five ways we can show we care at school. Which of these actions would you most appreciate?

1. **Active listening.** When someone is speaking, give them your full attention. Maintain eye contact and refrain from interrupting. Demonstrate empathy by acknowledging their thoughts and feelings, asking open-ended questions, and offering supportive and validating responses. By truly listening, you help others feel heard and valued.

2. **Offers of support.** Be proactive in helping your students and colleagues. Pay attention to their needs and challenges, and offer assistance when possible. This can be as simple as helping with a task, providing guidance or resources, or lending a listening ear. Demonstrate that you are willing to contribute to their success and wellbeing.

3. **Acts of kindness.** Small acts of kindness can go a long way. Take the time to recognise and appreciate your efforts and achievements. Offer words of encouragement, leave a thoughtful note or bring in treats to share. Simple gestures like these demonstrate your appreciation and create a positive and uplifting atmosphere at work.

4. **Respect and inclusion.** Show respect for other opinions, ideas and perspectives. Encourage open discussions valuing a diversity of viewpoints. Create an environment where everyone feels comfortable expressing themselves. Avoid judgment and actively promote collaboration and teamwork. Respecting others' contributions demonstrates that you value and care about their input.

5. **Celebration of milestones.** Take the time to honour achievements. Whether it's a work anniversary, a promotion or a personal accomplishment, your acknowledgment demonstrates that you are invested in the happiness and growth of others. Send a congratulatory message, organise a small gathering or show recognition in meetings.

2. Psychological safety and trust

Psychological safety means that we can take interpersonal risks such as expressing ideas, asking questions and making mistakes without fear of negative consequences. Trust means having confidence in the intentions, reliability and competence of others. Both create an environment where teachers feel supported and respected, leading to improved job satisfaction and wellbeing.

Trusting relationships foster a sense of support, collaboration and shared responsibility. Tschannen-Moran and Hoy (2001) found that teachers who perceive higher levels of trust in their colleagues and administrators experience lower levels of stress and greater job satisfaction. Psychological safety promotes risk-taking, creativity and innovation in the workplace. This allows for continuous growth, learning and professional development. A study by Edmondson (1999) revealed that teachers who feel psychologically safe are more likely to engage in reflective practices, share ideas and collaboratively solve problems.

Psychological safety and trust contribute to effective communication and conflict resolution. A study by König et al. (2016) demonstrated that teachers who perceived higher levels of trust engaged in more constructive communication, leading to better conflict management and a reduction of negative emotions. Feedback provided in a constructive and empathetic manner is key to this. When teachers feel safe to express themselves to their colleagues and administrators, they are more likely to seek and receive emotional support. The social network plays a crucial role in buffering the negative effects of stress and promoting wellbeing (Özbilgin et al., 2016).

→

OVER TO YOU

In Chapter 3 we were introduced to the concept of character strengths. For this activity you are encouraged to spot strengths in colleagues and share this feedback with them.

- Create a deck of cards with various character strengths written on them. After noticing a person using a particular strength, place the appropriate card in their pigeonhole. Be sure to add a small note that explains what you noticed and why you are sharing this feedback with them.

- As you move through your day, observe how people are using different strengths. Share your strength-spotting feedback in conversation with a colleague. Be sure to name the strength you saw them using with an example of how they used it. This can be a very positive and affirming activity for you and the other person.

3. High-quality connections

According to organisational psychology expert Jane Dutton, positive short interactions can form high-quality connections (Dutton and Heaphy, 2003). These interactions leave us feeling energised and invigorated. They can happen as we pass through the corridor or even in a five-minute parent meeting.

People who engage in high-quality connections at work are physically and psychologically healthier, with better cognitive functioning (Fredrikson, 2013). They tend to be more resilient when faced with challenges due to their higher level of engagement (Stephens, 2003). When energised by the people around us at work, we are more collaborative and open to learning new skills and hearing feedback.

Dutton and Heaphy (2003) have identified four pathways to high-quality connections:

1. **Respectful engagement.** This pathway involves actively listening to colleagues, expressing genuine interest in their ideas and perspectives, and treating them with kindness and dignity. Respectful interactions establish trust and psychological safety by fostering high-quality connections.

2. **Task enabling.** We can support others in their work-related tasks by offering help, sharing resources, providing guidance and collaborating effectively. By actively facilitating the success of others and demonstrating a willingness to contribute to their achievements, we enhance our high-quality connections.

3. **Trusting connections.** By establishing a sense of trust with colleagues we demonstrate integrity, dependability and resilience. When we consistently act in trustworthy ways, we contribute to the creation of high-quality connections based on reliability.

4. **Task-related play.** Engaging in enjoyable and creative interactions makes work more enjoyable, fosters a sense of fun and creativity, and encourages innovation and collaboration. By incorporating playfulness and creativity into work interactions, we can foster positive energy and build high-quality connections.

OVER TO YOU

Consider how you engage in high-quality connections by reflecting on the following four questions:

- How do you show kindness to colleagues?
- How do you help colleagues do their job well?
- How do you show that you trust your colleagues?
- How are you playful with colleagues?

4. Corrosive connections

Corrosive connections are negative interactions or relationships that have a detrimental effect on us and our work environment. These connections create toxic dynamics that lead to decreased productivity and job satisfaction. Schools need to promote open communication, provide training on respectful interactions, encourage collaboration and address any instances of bullying, harassment or toxic leadership promptly and effectively.

Examples of corrosive connections at work include:

- **Bullying and harassment.** Bullying or harassment from colleagues or superiors can include behaviours such as belittling, name-calling, intimidation and spreading rumours. These corrosive connections erode trust, create fear and can have severe psychological consequences for victims.

- **Excessive competition and sabotage.** In environments where competition is encouraged to an unhealthy extent, individuals may engage in sabotaging behaviours to gain an advantage: withholding information, undermining coworkers' efforts, engaging in covert tactics to ensure personal success at the expense of others. This weakens teamwork, fosters a culture of distrust and hinders collaboration and progress.

- **Negative leadership.** Leaders who exhibit negative traits such as micromanagement, favouritism and lack of empathy create corrosive connections with their teams. Employees may feel demoralised, undervalued or unappreciated, leading to decreased motivation, engagement and job satisfaction. Negative leadership styles can contribute to high turnover rates and a toxic organisational culture.

- **Gossip and rumours.** Spreading gossip or rumours creates an atmosphere of mistrust and undermines team cohesion. Gossip can harm professional relationships, damage reputations and breed a toxic work environment where individuals feel unsafe and vulnerable.

- **Lack of communication and collaboration.** Poor connections weaken relationships in teams and departments. Silos and barriers are formed, hindering the sharing of ideas, knowledge and resources. This lack of synergy can lead to inefficiencies, misunderstandings and a fragmented work environment.

OVER TO YOU

Sometimes corrosive connections can occur when we are blinkered to other people's perspectives. By taking a moment to stand in another's shoes, we give ourselves an opportunity to build empathy and understanding.

- Consider a recent situation in your teaching experience that involved a challenge, a conflict or a difference in viewpoints.

- Take a moment to imagine yourself in the other person's position. Consider their background, experiences, beliefs and emotions. Try to step out of your own perspective and fully immerse yourself in their situation.

- Reflect on what could be motivating the other person's actions or responses. What might their concerns, fears or desires be in the situation?

- Think about external factors such as cultural, social or personal circumstances that could be influencing the person's viewpoint. Consider how these factors contribute to their unique perspective.

- Reflect on the preconceived notions that may have influenced your understanding. How might these have limited your ability to see the situation from the other person's perspective? Be open to challenging and reevaluating yourself.

- Consider the lessons you've gained from this exercise. How has stepping into the other person's shoes changed your understanding of the situation? What new perspectives or considerations have emerged? Reflect on how this newfound understanding can inform your future interactions and decision-making.

- Based on your reflections, identify specific actions or strategies that you can implement to enhance your empathy in interactions. Consider how you can incorporate different perspectives and actively seek to understand other viewpoints in your teaching practice.

5. Seeking support

Being able to seek support is an important skill. Reaching out to trusted colleagues, mentors and friends allows us to alleviate a burden we are experiencing. For teachers, it's also valuable to seek support through collaboration. This can involve sharing challenges and brainstorming solutions together, co-planning lessons or seeking feedback on instructional strategies. By working collaboratively, we can alleviate stress and benefit from the wisdom and experience of our colleagues.

Teachers can find it hard to ask for help. Expectations of competence and capability can lead to a fear of appearing inadequate if we seek assistance. The demands of the profession can also make it difficult for us to prioritise our own needs. A lack of trust or support within the school environment may discourage us from reaching out. We may internalise the belief that asking for help is a sign of weakness, leading to a reluctance to admit our struggles and vulnerability.

OVER TO YOU

Let's explore avenues for seeking help when you are feeling overwhelmed.

- Brainstorm five ways you could seek collaborative support from a colleague. These could include sharing teaching materials, participating in grade-level or subject-area team meetings to share challenges and seek advice, or forming a community to discuss strategies and problem-solving.

- Consider two people you could approach for possible mentoring or coaching. Mentoring involves working with a person more experienced than yourself to help you grow, while coaching involves an advisor with specific skills to help you set and achieve goals.

- Research at least three ways you could access external support. These could include your school's Employee Assistance Program (EAP), a counsellor through your GP or a stress-management workshop.

6. Acts of kindness

Engaging in thoughtful and considerate behaviours promotes wellbeing and happiness. Kindness can manifest in many ways: an offer of support, an expression of gratitude, provision of assistance, a show of understanding, a demonstration of respect. It goes beyond superficial politeness to encompass genuine care and concern for others' needs and feelings. Kindness is a fundamental human value that strengthens communities and creates a more harmonious society.

Research has shown that kindness at work is associated with increased wellbeing, job satisfaction and happiness (Nelson and Lyubomirsky, 2012).

Kindness fosters a sense of camaraderie, trust and cooperation among colleagues. Kind and supportive behaviours create a positive team culture that promotes collaboration, effective communication and better problem-solving (Cameron et al., 2015).

Acts of kindness have been linked to lower stress levels and burnout among employees. Kindness from colleagues and supervisors can buffer the negative effects of work-related stress and contribute to resilience and wellbeing (Spreitzer et al., 2015). A kind and supportive work environment increases engagement and productivity. When employees feel valued and appreciated, they are more motivated and invested in their work (Dutton et al., 2014).

Promoting kindness at work contributes to the development of a positive organisational culture. Organisations that prioritise kindness and empathy are more likely to attract and retain talented employees, and they enjoy a positive reputation both internally and externally (Kaplan et al., 2019).

Surprisingly, research has shown that consciously choosing to be kind can benefit the giver more than the receiver because of the release of dopamine and oxytocin in the brain. It's also interesting to know that performing a 'random act of kindness' is more powerful than conducting the same act of kindness over and over. One reason that acts of kindness benefit our wellbeing is that they reinforce our view of ourselves as purposeful and generous.

OVER TO YOU

There are few better ways to express gratitude and appreciation to a colleague than with a heartfelt thank-you letter.

Reflect on a colleague who has made a positive impact on your professional life or provided support in some way. Think about the qualities you appreciate in this person and draft your letter. Consider tone, language and specific examples that best convey your appreciation. Here is an outline to help you structure your letter:

- **Opening.** Start with a warm and personalised greeting.
- **Express gratitude.** Clearly state the reasons for your appreciation and the actions or qualities that have made a difference.

- **Share impact.** Describe how your colleague's actions or qualities have positively affected their own work or the work environment.
- **Closing.** End with a sincere expression of gratitude and well wishes.

Share the letter in-person, by email or in an anonymous note left in your colleague's pigeonhole. I like to hand-deliver a card with a note inside.

After the letter has been shared, reflect on your experience of writing and expressing gratitude. How has this activity affected your wellbeing and work environment?

Impact

The value of acknowledgment

In the bustling world of education, it's all too easy for us to overlook the positive impact we have on others. As we will learn, the ability to notice and celebrate our multifaceted influence on the school community is vital for us to thrive.

Research has consistently highlighted the profound impact we have on our students. Positive teacher-student relationships are linked to increased academic achievement, motivation and overall wellbeing (Roorda et al., 2011; Wentzel, 2018). Our beliefs, attitudes and instructional practices can significantly affect student self-perception, resilience and lifelong learning orientations (Hattie, 2012; Pianta et al., 2020). Despite the weight of this influence, we often downplay or overlook the positive outcomes we create.

Acknowledging the difference we make bolsters our sense of purpose, self-efficacy and job satisfaction (Johnson et al., 2018). It provides validation and affirmation of our efforts, igniting a positive feedback loop that fuels our commitment to teaching. By nurturing an awareness of our influence and intentionally celebrating it, we can cultivate agency, purpose and joy in our teaching.

This chapter provides practical strategies for us to cultivate this mindset in our daily practice. From reflection exercises that unveil the ripple effects of our influence to collaborative activities that foster a culture of appreciation, we will gain insights and tools to amplify our impact and derive greater fulfillment from our profession.

Know thy impact

Coined by Professor John Hattie, this phrase emphasises the importance of understanding our influence on student learning and wellbeing. The concept aligns with Hattie's visible learning research, which emphasises the power of teachers to shape student outcomes (2009). When we are aware of our impact, we experience greater professional fulfillment and a sense of meaning in our work. We also become more attuned to the progress and achievements of our students. This awareness helps us recognise our strengths and areas for improvement, allowing us to refine our teaching practices and tailor our instruction to meet diverse student needs. Our resulting sense of agency and effectiveness leads to increased job satisfaction and reduced burnout.

Knowing our impact can significantly contribute to self-efficacy. When you recognise the positive influence you have on student learning, your belief in your ability to make a difference is reinforced. Observing your students' progress, witnessing their achievements and receiving feedback provides tangible evidence of your effectiveness as a teacher. This recognition reaffirms your belief in your efforts and reinforces your self-efficacy.

Understanding your impact connects you to purpose and meaning in your work. Seeing how your teaching strategies, guidance and support contribute to your students' academic and personal development strengthens your conviction that your role as an educator matters.

Recognising the time and energy you put into others can provide you with resilience and determination in the face of challenges. Drawing on positive changes you facilitated in the past can help you to overcome obstacles and persist in your efforts. This reflection serves as a reminder of your ability to navigate challenges, adapt your teaching strategies and find solutions. Your bolstered confidence will make you more likely to take risks, try innovative approaches and step outside your comfort zone.

Toolbox of activities

1. Reframe the belief of 'not enoughness'

Managing conflicting priorities is an ongoing challenge for educators who find themselves pulled in multiple directions. We face high expectations from administrators, parents, students and society. There is pressure to deliver high-quality instruction, facilitate student growth and meet academic standards. Struggling to meet these expectations within the constraints of time and resources can lead to feelings of inadequacy.

Let's not forget that we are deeply invested in our students' success and wellbeing. We have a strong desire to make a positive impact on their lives and provide them with the best education possible. This personal investment and passion can create a sense of responsibility and a drive to do more.

Researcher and author Brené Brown has extensively explored the concept of 'not enoughness' and its impact on our wellbeing and sense of belonging. She highlights that society often promotes a culture of scarcity, making us feel that we are not measuring up or doing enough (Brown, 2010). We thus set ourselves expectations that are unrealistic, unattainable and unhelpful. Our insecurities also stem from our desperate need to feel validated for our hard work. 'Not enoughness' feeds our feelings of lack, shame, fear and guilt.

My inner critic has always been loud. For years I blamed myself for not working hard enough. No matter how organised I was, I couldn't seem to get ahead. When I did pause, I guiltily remembered the lessons to be differentiated and the assignments to be marked. When I didn't meet my own deadlines and expectations, I felt shame. It was a vicious cycle.

There are two main triggers of 'not enoughness'. The first one is comparison. We may be proud of an achievement, only to realise that the teacher in the next classroom has done the same thing more successfully. Comparison can be crushing, but it can also be motivating. The second trigger is excessive expectations. If we're exhausting ourselves striving for unrealistic results that we consistently fall short of, it will only reinforce our belief that we are not enough.

Brown emphasises the need to develop a sense of self-worth independent of external achievements or validation. Recognising our inherent worthiness and practicing self-compassion allows us to embrace imperfections and be kind to ourselves. An important contributor to this is embracing our vulnerability. By acknowledging our struggles, we can cultivate authentic connections, courage and resilience.

Brown suggests that to counter the feeling of 'not-enoughness' we need to live wholeheartedly by embracing a sense of worthiness, courage and connection. Wholehearted living means cultivating gratitude, joy and resilience. It also requires us to set boundaries and practice self-care.

It's important that we set realistic expectations, seek support from colleagues and celebrate small victories. Teaching is a continuous process, so focusing on your positive impact can promote a healthier mindset.

➡️

OVER TO YOU

Brené Brown recommends a reflective activity called the 'hustle for worthiness inventory' to counter feelings of 'not-enoughness' and challenge the mindset of constantly striving for validation (2010). This activity encourages us to reflect on our beliefs and behaviours related to worthiness and offers an opportunity for self-exploration and self-compassion.

Reflect on the following questions and journal your responses:

- What are the ways in which you hustle for worthiness in your life? These could include external validation-seeking, perfectionism, overworking, people-pleasing or comparing yourself to others.
- How do these beliefs and behaviours impact your wellbeing, relationships and sense of worth?
- Are there any triggers or situations that intensify your feelings of 'not-enoughness'?
- What underlying beliefs or messages fuel your feelings of inadequacy or unworthiness?
- After reflecting on your responses, review what you have written with a compassionate and non-judgmental mindset. Practice self-compassion by acknowledging that these feelings and behaviours are common and understandable. Remind yourself that you are worthy of love and belonging simply by being human.
- Consider alternative beliefs and behaviours that can counteract the hustle for worthiness. These could include embracing self-acceptance, setting boundaries, practicing gratitude, cultivating authentic connections and prioritising self-care.
- Identify actions you can take to integrate these alternative beliefs and behaviours into your life.

2. Remember why you chose to be an educator

Michael F. Steger, a leading researcher in positive psychology, has extensively explored the importance of meaning in individuals' lives and its impact on their wellbeing. Steger contends that finding and cultivating meaning in one's work can greatly contribute to personal fulfillment and thriving.

Steger defines meaning as 'the significance that people ascribe to their lives and the purpose and direction they experience.' (2012) It goes beyond mere happiness or satisfaction to a deeper sense of purpose and coherence in life. Meaning can be derived from personal values, relationships, accomplishments and contributing to something larger than oneself.

Finding meaning in our work enhances our wellbeing and professional fulfillment. When we perceive our work as meaningful, we are more likely to experience higher job satisfaction, engagement and psychological wellbeing (Steger, 2012). As teachers, we feel a sense of purpose seeing our positive impact on our students and society as a whole. Meaning can help us navigate challenges and sustain our motivation. Research suggests that educators who find meaning in their work are more resilient, experiencing less burnout and higher levels of job satisfaction even in demanding circumstances (McMahon, Wernsing and Luthans, 2011). Connecting our daily tasks and efforts to a larger purpose can help us find greater fulfillment, maintain our enthusiasm and thrive in our professional lives.

OVER TO YOU

As we learned in Chapter 3, we can find meaning in our work by remembering our reasons for entering the teaching profession. Take a moment now to reconnect with your initial motivations. Reflect on the following questions:

- What inspired you to become a teacher? What were your aspirations?
- Think back to an experience that reaffirmed your passion for education. Describe its significance to you.
- How have you made a positive impact on the lives of your students? Recall stories that highlight the difference you make.
- What was most meaningful to you back when you were at school? What or who had a significant impact in shaping you?
- Reflect on the joy and fulfillment you experience when witnessing student growth, learning and success. Describe a memorable achievement or breakthrough that brought you deep satisfaction.

3. Start with your why

Created by inspirational speaker Simon Sinek, the Golden Circle model of human behaviour consists of three concentric circles: the *why*, the *how* and the *what* (Sinek, 2009).

Representing the core purposes and beliefs that drive us, the *why* encapsulates the deeper meaning and values behind our actions. The *how* refers to the strategies, processes or methods employed to fulfill that purpose. The *what* represents the tangible products, services or actions that result from the *why* and the *how*.

Embracing the Golden Circle can help us connect with our purpose so we can thrive. By starting with the *why* we can clarify and articulate our underlying motivations and core beliefs about education. This includes understanding the impact we want to have on students, the values we hold and the broader vision we seek to fulfill. A clear *why* provides us with direction and purpose. A team that is guided by a common *why* is united in camaraderie and collective responsibility.

➡️

OVER TO YOU

Take a moment to reflect on the meaning and impact of your work using Sinek's Golden Circle framework. By exploring your core motivations, aligning strategies with beliefs and examining the tangible outcomes of your teaching practice, you can gain deeper insights into your purpose and the influence you have on your students.

- Draw three concentric circles to represent the Golden Circle.
- Start from the centre circle, labelled *why*. Write down the motivations, beliefs and values that drive your work as an educator. Draw on your responses from the previous activity.
- Moving to the *how* circle, consider the strategies you employ to fulfill your core motivations. These could include instructional practices, classroom management techniques or pedagogical approaches.
- In the *what* circle, write down the tangible outcomes you see as a result of your teaching practice. These could include specific achievements, positive changes in students or broader contributions to the school community.

- After completing your reflections, what did you notice about the alignment between your core motivations and the strategies you use to make an impact?
- Create a personal commitment or action plan to further align your teaching practice with your core motivations and beliefs.

4. Reflect on your best possible self

This positive psychology intervention requires you to imagine your most fulfilled future self. Envisioning a positive and successful future will enable you to tap into your aspirations, strengths and goals.

Envisioning a positive future fosters optimism (King, 2001). This exercise has been found to enhance positive affect, life satisfaction and overall psychological wellbeing (King, 2001; Sheldon and Lyubomirsky, 2007). It can increase intrinsic motivation, goal-directed behaviour and efforts toward achieving personal goals (Sheldon and Lyubomirsky, 2006).

By focusing on positive possibilities, you may experience a reduction in negative thoughts and rumination (Lyubomirsky and Layous, 2013). You may also strengthen your resilience by using a positive framework to face challenges and setbacks (Sheldon and Lyubomirsky, 2006).

OVER TO YOU

Take a moment to reflect on your best possible self by doing the following:

- Close your eyes and take a few deep breaths to centre yourself.
- Envision a not-too-distant future when you will have achieved your goals and fulfilled your potential to live your most meaningful life.
- Visualise various aspects of this life: relationships, career, personal growth, health, wellbeing. Imagine yourself excelling and experiencing fulfillment in these areas.
- Engage your senses and emotions to imagine how it feels, looks, sounds and even smells to be your best possible self.
- After immersing yourself in this vision, open your eyes and write a detailed description of your best possible self. Capture your aspirations, accomplishments and values. Detail the positive emotions associated with your envisioned future. Be as descriptive as possible.

5. Seek strength-focused feedback

Feedback plays a crucial role in building self-efficacy by helping us believe that we can achieve desired outcomes and effectively perform our role. According to the late psychologist Albert Bandura's social cognitive theory, feedback serves as a key source of information that individuals use to evaluate their competence and adjust their behaviours (Bandura, 1997). Positive and constructive feedback provides us with valuable information about our instructional practices, effectiveness in meeting student needs and areas for improvement.

Research reveals that feedback related to teaching practices and student progress positively influences teachers' beliefs in their instructional competence and contributes to their sense of efficacy (Tschannen-Moran and Woolfolk Hoy, 2001). Teachers who receive frequent and specific feedback about their teaching practices demonstrate higher levels of self-efficacy than those who receive limited or no feedback.

Effective feedback is timely, specific and focused on strengths and areas for improvement. It should provide us with actionable information, support our professional growth, and align with our goals and needs. The problem arises when we take this feedback personally and see it as criticism. One approach that can mitigate undue offence is strength-based feedback that acknowledges talents and character. This feedback can help us develop a deeper understanding of our positive qualities and enhance our perception of professional competence (Clifton and Harter, 2003).

Focusing on strengths can increase intrinsic motivation and engagement among teachers (Lopez and Louis, 2009). A reinforced sense of competence energises us to excel in our work. This increased engagement contributes to improved job performance and student outcomes (Waters, Marzano and McNulty, 2003).

Strength-based feedback offers you insights into areas where you excel. This information can guide your professional growth and development by providing a foundation to build upon existing strengths (Waters et al., 2003). When teachers receive strength-based feedback, they in turn are encouraged to recognise the unique strengths of their colleagues (Rath and Conchie, 2008). This fosters a culture of collaboration and teamwork that leverages collective strengths to improve student outcomes (Rath and Conchie, 2008).

Strength-based feedback is not only beneficial to our professional growth but can also positively impact student learning. Research shows that teachers who focus on leveraging their strengths in the classroom create more engaging and effective learning environments (Hattie, 2009). When we recognise and use our strengths, we can tailor our instruction to better meet the needs of our students, resulting in improved academic outcomes.

→

OVER TO YOU

In this activity you are encouraged to be brave and seek feedback from either your colleagues or your students about the strengths they see in you. You will gain insights into your positive qualities and how they impact others. This activity fosters a culture of appreciation, collaboration and continuous growth.

- Approach your intended audience about a strength-appreciation session you would like to engage in. Explain the purpose of seeking feedback on your strengths and positive qualities as a way of nurturing what you do well.

- Provide a feedback form that includes prompts to highlight your strengths. The questions can ask about moments when your strengths were observed, specific strengths you possess or examples of the positive impact you've had on others.

- You may wish to include the option to provide anonymous feedback. This can encourage honest and open responses.

- The next step is to review the feedback you've received. This must be done in a supportive and constructive manner. You can choose to have colleagues or students take turns reading out the feedback they provided, or you could opt to read it privately.

- Reflect on the strengths that were highlighted and consider how they align with your self-perception. Do you see your own strengths or is your inner critic drowning them out?

- Identify ways to leverage your strengths in your teaching practice. You might incorporate them into lesson planning, explore new teaching strategies or share your expertise with colleagues.

6. Building self-efficacy

High self-efficacy makes us more likely to set goals, with less fear of failure and greater willingness to apply effort under stress (Artino, 2012). When self-efficacy is low, we tend to avoid tasks. Both children and adults with high self-efficacy tend to be optimistic about the future and have higher levels of mental and physical health (Seligman, 2006).

Teachers with confidence in their professional abilities feel more control over their daily tasks (Tschannen-Moran et al., 2007), are more motivated (Skaalvik and Skaalvik, 2007) and experience greater job satisfaction (Hoy et al., 2005). These teachers successfully meet the rising demands of students, are willing to stretch themselves and grow, persist in the face of challenges, foster learning autonomy and convey high expectations (Donohoo, 2017).

We feel pride in ourselves as professionals when we are able to notice our positive impact (Buonomo et al., 2019). These positive emotional states can have an 'undoing effect' to restore psychological resources in the face of negative emotional states such as stress (Gloria et al., 2013).

➜

OVER TO YOU

Let's explore four ways to build our self-efficacy as teachers:

1. Mastery experiences

Practicing a task over and over builds mastery. When we perform a task successfully, it builds our confidence. This means setting clear goals and being prepared to make mistakes and learn from them.

- Describe instances when you felt successful or accomplished in your teaching practice. What did you do well? How did it impact your confidence?

- How can you intentionally create more opportunities for mastery experiences in your teaching? What goals can you set to further develop your expertise?

2. Social modelling

Seeing people similar to ourselves succeed through sustained effort raises our beliefs that we possess the capabilities to master comparable activities.

- Identify real-life people who serve as role models for you. What qualities do you admire in them?
- How can you learn from these people? What actions or strategies can you learn from?

3. Social persuasion

Receiving verbal encouragement helps us overcome self-doubt and focus on giving our best effort to the task at hand.

- Reflect on words of encouragement you've received from students, colleagues or administrators. What affirmations have stuck with you?
- How can you cultivate supportive relationships and feedback mechanisms to bolster your self-efficacy?

4. Emotional and physiological state

Our moods, emotions, physical state and stress levels all affect how we feel about our personal capabilities. We can improve our sense of self-efficacy by learning how to minimise stress and elevate mood when facing challenging situations.

- Reflect on your emotional and physical reactions in various teaching situations. How do certain emotions and sensations affect your confidence?
- How can you proactively manage stress, anxiety and other negative emotions to maintain a sense of control and confidence in your teaching?

Keep in mind that self-efficacy is not fixed, and can be strengthened over time. I encourage you to regularly revisit these sources of self-efficacy, and to intentionally seek out experiences and sources of support that nurture your confidence.

Values

The power of values

In the vast tapestry of education, teachers occupy a pivotal place. It is up to us to ignite the flames of knowledge, curiosity and growth within our students. We are the torchbearers of inspiration shaping the minds of future generations. But what fuels our pursuit of excellence? What drives us to overcome challenges, adapt to changing landscapes and find fulfillment in our dynamic profession?

We are empowered to thrive when our work closely aligns with our core values, the fundamental beliefs and principles that guide our thoughts and actions. Our core values represent the qualities and ideals that we hold dear. They are the guiding principles that shape our decision-making, inform our actions and establish the foundation upon which we build our professional lives.

Understanding our values helps us make decisions in alignment with our sense of purpose. When we integrate our values and our teaching practice, remarkable transformations occur. We become more focused, purpose-driven and resilient, unlocking our true potential.

This chapter explores the profound influence of values on a teacher's journey, illuminating the ways in which self-awareness and intentional alignment enhance professional growth and wellbeing. Let's discover the importance of clarifying personal values that can guide decision-making and help us set SMARTER goals.

Toolbox of activities

1. Identify your core values

Our core values serve as a compass that provides a sense of direction and purpose in our lives. Alignment with our core values fosters congruence and integrity. By continuously striving to embody our core values, we can learn from our experiences and expand our potential.

OVER TO YOU

Find a quiet and comfortable space where you can engage in self-reflection without interruption. Begin by brainstorming a list of values that resonate with you. Use the list below as a guide and feel free to add your own.

Write down any words or phrases that come to mind when you think about what is most important to you as a teacher. Focus on the qualities and principles that guide your actions, decisions and interactions with others. Allow yourself to be open-minded and non-judgmental during this process.

Core values

Authenticity	Achievement	Adventure
Authority	Autonomy	Balance
Beauty	Boldness	Compassion
Challenge	Citizenship	Community
Competency	Contribution	Creativity
Curiosity	Determination	Fairness
Faith	Fame	Friendships
Fun	Growth	Happiness
Honesty	Humour	Influence
Inner harmony	Justice	Kindness
Knowledge	Leadership	Learning
Love	Loyalty	Meaningful work
Openness	Optimism	Peace
Pleasure	Poise	Popularity
Recognition	Religion	Reputation
Respect	Responsibility	Security
Self-respect	Service	Spirituality
Stability	Success	Status
Trustworthiness	Wealth	Wisdom

Review your list of brainstormed values and highlight the five that resonate most strongly with you. Take some time to explore each value individually. Reflect on what that value means to you personally and how it aligns with your role as a teacher. Look for any common threads or connections between the values.

Ask yourself reflective questions about each prioritised value:

- What does this value look like in my teaching practice?
- How does this value affect my decision-making process?
- In what ways can I embody and promote this value in my classroom and the school?
- How do I feel when this value is present or absent in my work?

2. Connect to your inner compass

When we align our actions with our values, we experience the authenticity and purpose that come from living in accordance with what truly matters to us. Values enable us to make choices that are in harmony with our beliefs and aspirations. The following activity can serve as a guide for self-reflection and intentional decision-making.

OVER TO YOU

Begin by taking a few moments to focus on your breath. Close your eyes, inhale deeply through your nose and exhale slowly through your mouth. Allow your breath to bring you into a state of calm and centredness. Repeat this process several times, allowing any tension to melt away.

Take a sheet of paper and divide it into two columns. In the left column, list the values you identified as most important to your teaching practice.

In the right column, write down reflective questions that will guide your exploration of each value. These questions should encourage you to bring your values to life in your daily interactions and decision-making as a teacher. For example:

- How does *compassion* manifest in my interactions with students and colleagues?
- In what ways can I infuse *integrity* into my teaching practices?
- How do I foster *curiosity* and a *love of learning* in my students?
- How can I promote *collaboration* and *teamwork* in my classroom?

These actions can range from small adjustments in daily routines to more significant initiatives. Write down your action plan and commit to sharing it with a trusted colleague, mentor or friend who can provide support. Set regular intervals to review your progress, celebrate successes and make adjustments.

3. Set SMARTER goals

Do you know how to set goals effectively? Dr Edwin Locke argued that clear goals and appropriate feedback are a major source of motivation that improve employee performance (1968). He found that goals that were specific and moderately challenging led to higher performance than vague or easy goals. Later research also emphasises the need for goals to involve sufficient task complexity and to encourage commitment and feedback (Locke and Latham, 1990).

In 1981, George T. Doran created the SMART acronym to help us create goals that are specific, measurable, assignable, realistic and time-related. Today SMART has many variations that are used across numerous fields.

My version of SMART is adjusted for a teaching context, and becomes SMARTER with the addition of 'evaluate' and 'reflect'. After all, the purpose of goal-setting is not simply to achieve a specific outcome. It is also to reflect on our experience and apply what we have learned to similar events in the future.

Specific	Measurable	Authentic	Resources	Timely	Evaluate	Reflect
S	M	A	R	T	E	R
Choose a focus area	Make it measurable	Make it authentic	Plan your resources	Set an end date	Assess the outcome	Reflect on learnings

Core values play a significant role in setting SMARTER goals, as they provide the foundation and guiding principles for our objectives. Use your values as a filter for goal-setting. They will help you determine whether a particular objective is relevant and meaningful to you. Goals that align with your values are more likely to drive satisfaction and fulfillment.

When you have a clear understanding of your values, you can assess which goals are most closely aligned and prioritise accordingly. This helps you focus your efforts on the goals that truly matter to you. When faced with multiple options or opportunities, referring to your core values can help you make choices that are consistent with your beliefs and aspirations.

OVER TO YOU

Take a moment to think about how you could be SMARTER in setting meaningful and authentic goals. Use the framework and questions below to set yourself a goal that relates to your wellbeing at work.

	Question	Sample response
Specific	What area of your wellbeing would you like to focus on? Perhaps one of the PERMAH pillars?	*The area of wellbeing that I would like to develop is feeling valued.*
Measurable	What measures could you use to track your progress?	*I could use a weekly journal and ask for feedback from 3 people.*
Authentic	Why is achieving this important to you?	*I work very hard, yet don't feel rewarded for this work. This doesn't mean I'm being egotistical; it means I'm human.*
Resources	What resources do you need to help you achieve this? (People, time, technology etc.)	*I need to buy a journal, develop a feedback template and be brave enough to speak to 3 people.*
Timely	When would you like to complete this?	*I would like to do this by the end of the month.*
Evaluate	How will you evaluate the effectiveness of your efforts?	*Hopefully I will be able to quieten my inner critic and be prouder of my achievements.*
Reflect	Where will you record successes and areas for improvement?	*My journal.*

Now use your responses to craft a wellbeing goal. For example:

By the 31st of March, I will have received strength-based feedback from three different people. I will have recorded my experience of this in my journal with a focus on appreciating my good work.

4. Nurture your desire to feel valued

Although teachers are the pillars of education, we find ourselves grappling with feelings of being undervalued. This can have a profound impact on our self-worth, contributing to a range of psychological challenges. Wouldn't it be wonderful to see society, parents and even policymakers appreciate the dedication, expertise and personal investment we bring to our classrooms?

The reality is that trying to be everything to everyone can lead us to undervalue ourselves. Learning to say 'no' at work can significantly increase our sense of feeling valued. It is essential to establish clear boundaries within and between our personal and professional lives. By honouring your needs, you send a message that your time and wellbeing are valuable. This in turn fosters your sense of self-worth.

By saying learning to say 'no', we can focus on quality over quantity. This allows us to exercise professional autonomy and increases our ability to deliver high-quality education. By setting boundaries, you can assert yourself as a professional who values their own time and expertise. As others recognise and appreciate your ability to make thoughtful decisions and prioritise your work, your sense of value is strengthened.

Learning to say 'no' can be difficult. Our passion for helping and our dedication to student success provoke a strong sense of responsibility. We may fear disappointing others or being perceived as uncooperative. The hierarchical nature of the education system can also create a power dynamic that makes saying 'no' uncomfortable. These factors make it difficult to set boundaries and prioritise our wellbeing.

➔

OVER TO YOU

Take a moment to think about how you could establish boundaries around the amount of work you choose to take on.

Here are five different ways you can say 'no':

1. Direct and assertive approach

I appreciate your considering me for this task, but unfortunately I'm unable to take on additional work at the moment. I need to focus on completing my current responsibilities to the best of my ability.

2. Suggesting an alternative

I understand the importance of this project, and while I can't commit to taking it on entirely, I can offer some guidance or support to help someone else who may be available. I believe it's important to share the workload effectively.

3. Explaining prior commitments

Thank you for thinking of me for this task. Unfortunately, I have several ongoing commitments that require my attention. I want to ensure that I can give my full dedication to those projects, so I won't be able to take on anything additional at this time.

4. Setting boundaries and negotiating

I'm happy to help, but I have my hands full with my current workload. If you could provide more information about the urgency and priority of this task, we might be able to reevaluate and discuss which responsibilities can be adjusted or delegated to accommodate it.

5. Suggesting delegation to others

I'm honoured that you trust me with this responsibility, but I believe it would be better suited to someone with expertise or availability in that particular area. I can help identify or recommend someone who would be a great fit for this task.

It's important to be respectful, confident and honest when refusing someone. By offering alternative solutions, explaining your limitations or proposing delegation, you demonstrate your willingness to be a team player while still valuing your own boundaries and priorities.

5. Savour the good stuff

In positive psychology, 'savouring' refers to the intentional and conscious appreciation of positive experiences, thoughts and feelings. This could mean reminiscing about pleasant memories, relishing present experiences or anticipating future positive events. An understanding of our values supports our ability to savour positive emotions. When we are aware of what matters to us, we can align our experiences and actions with those values.

Savouring has been found to amplify positive emotions, leading to greater happiness and life satisfaction. By focusing on and intensifying positive experiences, we can enhance our emotional wellbeing (Bryant and Veroff, 2007; Quoidbach et al., 2010). Savouring encourages us to be fully present and engaged in the current moment. It promotes mindfulness and helps us appreciate and enjoy the richness of our experiences, leading to increased wellbeing (Kiken et al., 2015; Hanley et al., 2020).

Savouring can build resilience. By intentionally focusing on what matters to us, we can find strength and positive emotions even in difficult situations (Bryant et al., 2005; Hurley et al., 2016). Collectively savoured experiences foster social connections and promote positive relationships. By savouring a moment together, we can create meaningful bonds (Algoe et al., 2017; Rimé et al., 2019).

Savouring encourages us to cultivate appreciation of the positive aspects of our lives. By helping us recognise and value the good things, it contributes to a greater sense of wellbeing (Emmons and McCullough, 2003; Bryant and Veroff, 2007). By actively incorporating savouring practices into your everyday life, you can reap the benefits of increased positive emotions, mindfulness, resilience, social connections and gratitude.

There are many ways to savour positive emotions: mindful awareness, practicing gratitude, sharing positive experiences with others. Visual reminders such as photographs, artwork or symbols can be a powerful tool for savouring memories. You can display these reminders in your environment as a visual cue. I'm a fan of photography because it's simple and easy to organise for future viewing. If you want to get very fancy, you can even create your own posters or books for printing.

➡️

OVER TO YOU

Photography is a powerful tool for evoking emotions and intensifying connections. The purpose of this activity is to encourage you to use photography as a means of capturing and savouring positive experiences.

- Take a moment to go through your photos. Choose an image of an event that brought you joy or fulfillment—a personal achievement, a beautiful moment in nature, an interaction with loved ones.

- If you do not have any suitable pictures, take the next week to capture photos of moments that make you feel good.
- Create an album on your device and copy these photos into it. Be sure to give your album an exciting and uplifting name.
- Take a moment to really look at these photos to enhance the positive emotions associated with the experiences.
- Write in your journal about the emotions, thoughts and memories associated with the photos. Describe how you felt during the experiences and the significance they hold for you.
- You may like to create a collage or book that can serve as a visual reminder of your positive experiences. You might even decide to revisit a location depicted in the photos.
- Keep your photo album handy for whenever you need a boost of positivity.

By engaging in this activity, you can develop a habit that promotes mindfulness, gratitude and an increased awareness of the positive moments that enrich your life.

6. Mindfulness

As we learned in Chapter 4, mindfulness means 'paying attention, in a particular way, on purpose, in the present moment, nonjudgmentally' (Kabat-Zinn, 1990). It has been shown to have numerous benefits for wellbeing, and is especially useful for teachers who face unique stressors in their profession. It's important to note that mindfulness is a skill that takes time and practice to develop. Consistency and regularity are key to experiencing its benefits.

The following mindfulness techniques have been widely studied and recognised as effective methods.

- **Mindful breathing**. This technique focuses our attention on the sensation of the breath through observation of inhalation and exhalation patterns. Anchoring awareness in the breath brings us back to the present moment whenever the mind starts to wander. Research has shown that mindful breathing can enhance attention and reduce stress (Arch and Craske, 2006).
- **Body scan**. A body scan systematically directs our attention throughout the body, allowing us to notice sensations and any areas of tension or

discomfort. This technique promotes bodily awareness and helps us develop a more grounded and accepting relationship with our physical experiences. Studies have demonstrated that body-scan practices can reduce anxiety and enhance self-compassion (Grossman et al., 2004).

- **Loving-kindness meditation**. This practice involves cultivating feelings of love, compassion and goodwill towards oneself and others. Silently repeating affirming phrases or visualising loved ones can generate feelings of warmth and kindness. Loving-kindness meditation has been associated with increased positive emotions, empathy and overall wellbeing (Fredrickson et al., 2008).

- **Mindful eating**. This involves bringing mindful awareness to the process of eating through attention to the sensory experiences, tastes and textures of food. By slowing down and savouring each bite, we can develop a greater appreciation for the nourishment that food provides and cultivate a healthier relationship with eating. Research suggests that mindful eating can improve eating behaviours, reduce binge eating and enhance self-regulation of food intake (Kristeller and Wolever, 2011).

- **Mindful arrival**. Before entering the classroom, take a moment to pause outside the door. Close your eyes, take a few deep breaths and bring your attention to the present moment. Set an intention for the day—for example, to be fully present with your students and foster a positive learning environment.

- **Mindful movement**. Incorporate mindful movement practices such as yoga or stretching into your routine. Engage in gentle movements while maintaining awareness of your body and breath. This can release tension, improve circulation and promote relaxation, helping to alleviate stress.

- **Mindful time management**. Use mindfulness to manage your time effectively and prioritise tasks. Before starting a new activity, take a moment to check in with yourself. Assess your energy levels, focus and priorities. Set realistic goals and break tasks into smaller manageable steps. By approaching your work mindfully, you can reduce feelings of overwhelm and increase productivity.

- **Mindful reflection**. Engage in regular reflection to process and manage work-related stress. Set aside time to journal, meditate or simply sit quietly and reflect on your experiences, thoughts and emotions. This mindful reflection can help you gain insights, cultivate self-awareness and develop a more compassionate and supportive relationship with yourself.

- **Mindful boundaries.** Establish clear boundaries between work and your personal life by practicing mindfulness. When you're off-duty, be fully present and engage in activities that replenish your energy. By setting boundaries and being mindful of your needs, you can prevent burnout and create a healthier work-life balance.

OVER TO YOU

One activity that you can engage in to practice mindful eating is the Raisin/ Chocolate Exercise. This well-known practice based on teachings by stress-reduction expert Jon Kabat-Zinn is commonly used in mindfulness-based interventions. It brings awareness to the act of eating a single raisin or piece of chocolate, emphasising the sensory experience and cultivating a deeper connection with the present moment.

Begin by holding the raisin or piece of chocolate in your hand. Take a moment to observe its appearance, texture and shape. Notice any thoughts or expectations that arise.

- **Engage the senses.** Bring the raisin or piece of chocolate close to your nose and take a few deep breaths, noticing any scents. Then slowly explore its texture with your fingers.

- **Slow and deliberate eating.** Bring the raisin or piece of chocolate to your lips. Place it in your mouth without chewing it. Focus on the taste and texture as it rests on your tongue.

- **Chewing mindfully.** Begin chewing very slowly and intentionally. Pay attention to the sensations of chewing—how it feels against your teeth, the changing texture as it softens and the flavours that emerge.

- **Mindful swallowing.** When you feel ready, prepare to swallow the raisin or piece of chocolate. Notice any impulses to swallow immediately and pause for a moment. Pay attention to the movement of swallowing, the sensations as the raisin or chocolate moves down your throat, and the moment when it disappears.

Throughout this exercise, it's important to maintain an attitude of non-judgemental curiosity, observing your experiences without evaluating or criticising them. The purpose is to cultivate a heightened awareness of the present moment and the sensory experience of eating.

This exercise can help you develop a greater appreciation for the process of eating, improve your ability to recognise hunger and satiety cues, and create a healthier relationship with food.

Emotions

Emotions and meaning

Teachers navigate a complex web of responsibilities. We juggle curriculum planning, classroom management, student engagement and assessment, all while striving to create an environment that fosters growth and learning. Amid these demands, it is easy for us to neglect our own emotional wellbeing.

A strong focus on meeting the needs of students leaves little room for self-reflection and self-care. What many educators fail to recognise is that their emotional wellbeing is not simply a personal matter; it directly impacts their effectiveness and fulfillment in the classroom.

Emotions are psychological and physiological experiences that arise in response to internal or external stimuli. Complex and multifaceted, they play a fundamental role in our daily lives by shaping our perceptions, thoughts, behaviours and wellbeing. Emotions are rooted in the intricate interplay between biology, cognition and social environment. They are influenced by the activity of the brain, in particular the limbic system which includes structures like the amygdala and hippocampus.

Beliefs, memories and past experiences all have a role to play in our emotions. Social factors such as our interactions with others, cultural norms and societal expectations also influence their range and expression. We learn emotional expressions, regulation strategies and social cues through socialisation, which shapes our experiences and responses to a great extent.

This chapter delves into the crucial role of emotions in a teacher's life and underscores the necessity of emotional regulation for us to thrive in an educational setting. We will explore six activities that can empower you to cultivate emotional intelligence, enhance self-awareness and foster resilience. By understanding emotions, you can create a positive ripple effect that transforms the educational experience for students and the broader school community.

Toolbox of activities

1. Building emotional literacy

We all know that a school day can be an emotional rollercoaster. Our hearts and minds are lifted by meaningful moments, only to be crushed by heavy workloads. Just as we reach one deadline, we see another looming. We manage to climb through paperwork before systemic changes force us to adapt yet again. On some days we scream with excitement. On other days we scream in frustration. There is a lot that fills us with energy, and a lot that can bring us down.

Put simply, *e-motions* are energy in motion. They are made up of a subjective component (how we experience the emotion), a physiological component (how our bodies react to the emotion) and an expressive component (how we behave in response to the emotion). These components interact with each other to motivate us toward action. Being able to recognise and name our emotions is an important step in managing them. When emotions are labelled, we gain better control over our responses.

We should also aim to recognise and understand the emotions of others. This builds social awareness and stronger connections. Accurate emotional labelling facilitates effective communication and enhances conflict-resolution skills. Communicating our needs and concerns while recognising the emotions underlying the conflicts leads to more constructive and empathetic solutions.

One model that can help us identify emotions is the Wheel of Emotions developed by psychologist Robert Plutchik. This theoretical framework illustrates the relationship between primary emotions and their intensities. Plutchik proposed that these primary emotions combine and interact to form more complex secondary emotions (1980).

The Wheel of Emotions provides a visual representation of the wide spectrum of emotions and their relationships, aiding in the recognition and understanding of complex emotional experiences. We can use this wheel to pinpoint the core emotions we are experiencing and assess their intensity levels, facilitating self-awareness and emotional regulation.

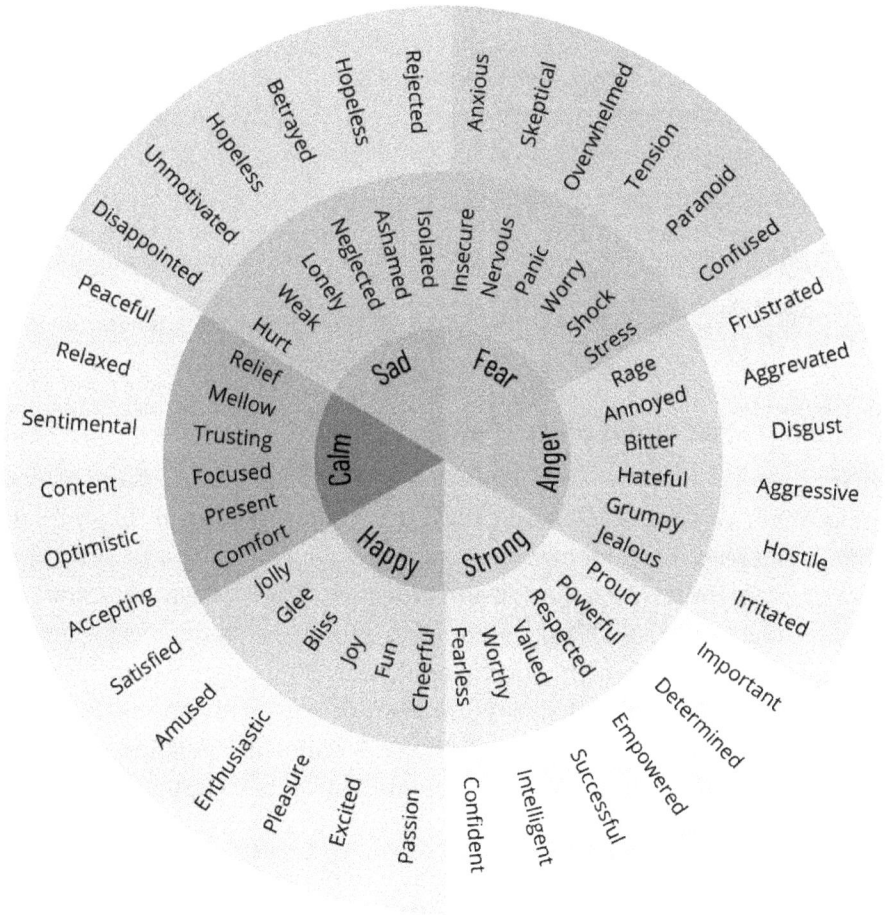

OVER TO YOU

Use the Wheel of Emotions to explore the emotions you typically experience in a day. Notice how these emotions are interconnected and may combine to form complex secondary emotions.

- Recall a recent emotional experience. Try to be as specific as possible in labelling the emotions you felt.
- Consider the intensity of each emotion. If you experienced fear, was it slight unease or full-blown panic?
- Explore the nuances of your experience to see whether any secondary emotions arose from the primary emotions you identified. For instance, did your fear also involve elements of surprise or anticipation?
- Write down your reflections. Describe the situation, the primary emotions you experienced, their intensities, and any secondary emotions that emerged. Reflect on how this affected your thoughts, behaviours and experience of the situation.

2. Managing emotions and triggers

Teachers who manage their emotions effectively create a positive and supportive classroom climate. They are better equipped to regulate their responses to student behaviour, providing a calm and nurturing environment for learning (Sutton and Wheatley, 2003). This fosters stronger student-teacher relationships, leading to increased student engagement and academic achievement (Curby et al., 2013). Teachers who are skilled in managing their emotions are more likely to cope with stress effectively, reducing the risk of burnout (Abel, 2019). Emotional regulation strategies such as self-care and stress management contribute to optimal wellbeing (Hascher and Waber, 2017).

As teachers, we serve as role models in emotional intelligence. When we demonstrate emotional self-awareness, regulation and empathy, we provide students with valuable examples of healthy emotional expression and conflict resolution (Brackett et al., 2011). This helps our students to develop their own emotional intelligence and social skills.

Emotional management also enables teachers to communicate effectively, especially in challenging situations. By understanding and controlling our emotions, we can respond thoughtfully. This leads to more productive and respectful interactions with students, colleagues and parents (Brackett et al., 2019). The ability to manage emotions reduces the likelihood of emotional exhaustion and disillusionment, contributing to long-term career fulfillment (Kyriacou and Chien, 2004).

There are helpful and harmful ways of dealing with our emotions, and the key here is self-awareness. Some of us will choose to run to the park and others will choose to run to the fridge. Some will sit quietly and focus on deep breathing, while others will choose to drown feelings of overwhelm or exhaustion in a bottle of booze. The activities in this chapter are designed not to judge your actions, but to raise your awareness of their consequences.

Let's explore positive and negative emotional triggers, and reflect on how we respond when faced with them. Positive triggers include student success, meaningful interactions and personal growth. Negative triggers include student challenges, workload pressure and a lack of support.

OVER TO YOU

Take a moment to reflect on your level of emotional intelligence and create a plan for better managing your emotions.

- Begin by reflecting on your emotional experiences this week. Consider one situation where you faced a challenging emotion and another where you experienced a positive emotion.

- Identify the emotions you experienced in these situations. Use descriptive language to accurately capture the essence of the emotion. Instead of simply labelling an emotion as 'anger', you could specify it as frustration, irritation or resentment.

- Reflect on the triggers that led to the emotion. What specific events, thoughts or interactions contributed? Understanding these triggers can help you become more aware of the patterns or situations that tend to elicit certain emotions.

- Evaluate the intensity of the emotion on a scale of 1 to 10. This assessment helps you gauge the potential of the emotion to influence your thoughts and behaviours.

Let's look at the negative emotion that you identified and explore how you responded to it.

- Which of your underlying needs or values were not met in the situation? Emotions often arise when our core needs are not fulfilled.

- Think of alternative ways of responding to the situation that would have aligned with your needs and values. Brainstorm strategies to help you manage similar emotions in the future. These might include deep breathing, cognitive reframing, seeking social support, practicing self-care or engaging in problem-solving.

- Develop an action plan to incorporate the identified emotion-management strategies into your daily life. Set achievable goals and establish a timeline for implementation. Consider how you can integrate these strategies into your routine to proactively manage emotions in future situations.

- Regularly reflect on your progress in implementing the action plan and evaluate its effectiveness. Adjust and refine your strategies as needed. Embrace a growth mindset and be open to learning.

Be kind to yourself throughout this process. Managing emotions is an ongoing journey that takes time and practice. Embrace self-compassion as you navigate your emotions and work towards greater emotional intelligence.

Below are some examples of emotion-provoking situations that you might encounter at school.

Triggers of POSITIVE emotions	Name the emotion	Intensity (1–10)	My response	Is this helpful or harmful?
Staffroom banter with colleagues	Joy	8	Laughing uncontrollably	Helpful
Student grades improving	Pride	5	Savour a moment to feel my heart expand with pride	Helpful
Positive feedback from a parent	Surprise	4	Brush off the positive comment	Harmful

Triggers of NEGATIVE emotions	Name the emotion	Intensity (1–10)	My response	Is this helpful or harmful?
Being asked to take on extra playground duty	Frustration	7	Huff, puff and storm out of the staffroom	Harmful
A student becomes argumentative	Anger	8	Focus on my breathing before responding	Helpful

3. Closing the stress cycle

In Chapter 2, we spoke about understanding the difference between stressors and the stress response, and the importance of closing the stress cycle. When the stress cycle remains open or unresolved, the body can experience prolonged activation of the stress-response system. This chronic stress response can have detrimental effects on our nervous system and overall health.

Here are five fun and easy ways that teachers can close the stress cycle:

1. **Mindful breathing.** Engage in a short mindfulness practice by closing your eyes, taking slow deep breaths and focusing on the sensations of breathing. This simple practice can help calm the mind and release tension.

2. **Expressive writing**. Spend a few minutes journaling about your day's experiences. Write freely without judgment to express your thoughts and emotions, as well as any challenges or successes. This process can help you to release stress and gain perspective.

3. **Physical activity break**. Engage in a quick burst of physical activity. Take a walk around the school grounds, do some stretching exercises or dance to a favourite song. Physical activity promotes endorphin release, boosts mood and reduces stress.

4. **Laughter therapy**. Watch a funny video, read a humorous book or share jokes with colleagues. Laughter triggers the release of feel-good hormones, providing an instant stress relief and creating a positive atmosphere.

5. **Creative outlet**. Engage in a creative activity that brings you joy. It might be drawing, painting, playing a musical instrument, engaging in crafts. This allows for self-expression and diversion from work-related stressors.

Find activities that resonate with your personal interests and preferences. These enjoyable practices help us shift our focus away from stressors, activate the relaxation response and close the stress loop.

→

OVER TO YOU

Now that we've explored ideas for closing the stress cycle, it's your turn to think about what would work best for you. Let's start with an example.

Strategy	Likelihood of doing this (1–10)	What this looks like for me	When I could do this	How could it make me feel?
Laughter	9	Watching a silly cat video for 3 minutes	At 3.30pm in the staffroom at the end of a school day	Lighter and happier
Mindful breathing				
Expressive writing				
Physical activity				
Laughter				
Creativity				

4. Acknowledging vulnerability

Brené Brown has extensively studied vulnerability and its impact. She emphasises the importance of embracing vulnerability as a necessary component of a wholehearted and authentic life, arguing that it is a sign not of weakness but rather of courage and strength (Brown, 2012). Brown suggests that vulnerability is a prerequisite for meaningful connections and a sense of belonging. By allowing ourselves to be vulnerable and authentic, we open the door to deeper and more fulfilling relationships with others (Brown, 2010).

Brown proposes that vulnerability is intertwined with courage and resilience. It takes courage to take risks despite our fear of rejection or failure; embracing this vulnerability can lead to increased resilience in the face of adversity (Brown, 2015). Vulnerability can also fuel creativity and innovation: by stepping into the discomfort of uncertainty, we unlock our creative potential and discover new possibilities (Brown, 2010).

Teachers experience vulnerability throughout our workday. It could be provoked by the challenges faced in delivering a lesson, or by the difficulties of managing student behaviour. Communication with parents can also make us feel exposed and vulnerable to judgment or conflict (Abidin, 2012). Then there's the vulnerability we can feel when receiving constructive criticism, which may trigger self-doubt and feelings of inadequacy (Nguyen et al., 2016).

Vulnerability can trigger feelings of insecurity, self-doubt and a diminished sense of self-worth. We may even question our competence, expertise and value as professionals (Brown, 2010). This can negatively affect our confidence, motivation and job satisfaction (Hauge, 2019). Even opening up about our struggles or vulnerabilities can create a sense of exposure and potential self-doubt (Kaplan et al., 2018).

In her book *The Power of Vulnerability* (2012), Brown explains various ways that we respond to vulnerability:

- **Armouring.** This is a defence mechanism we employ to protect ourselves from emotional exposure. It can manifest as being overly guarded or adopting a persona that presents a strong and invulnerable exterior. Armouring prevents us from fully engaging with vulnerability and connecting with others authentically.

- **Numbing.** We might avoid or suppress uncomfortable emotions by engaging in behaviours or distractions that help us escape vulnerability. Numbing can take the form of overworking, substance abuse and other unhealthy coping mechanisms.

- **Perfectionism.** This response is driven by the fear of being judged. It involves striving for flawlessness and placing immense pressure on oneself to avoid exposing weaknesses. Perfectionism can prevent us from taking risks or embracing vulnerability lest we encounter failure or.criticism.

- **Control.** The desire for control is a response to vulnerability and uncertainty. We may attempt to regain a sense of control by micromanaging situations, people or outcomes. The need for control can stem from the fear of not knowing how things will unfold.

- **Overachieving.** We may seek validation and worthiness by constantly striving for success, recognition or external accomplishments. By overachieving, we believe that we can shield ourselves from feeling vulnerable or being judged.

OVER TO YOU

Now that we've explored helpful and harmful ways of managing vulnerability, it's time to reflect on your own strategies:

- Take out a journal or a piece of paper and divide it into two columns.

- In the first column, make a list of recent situations where you felt vulnerable personally or professionally.

- Reflect on whether your response to each situation was typical of how you manage vulnerability. Be honest and observe your behaviours. Consider whether you tend to armour up, numb your emotions, seek control or engage in other responses.

- In the second column, write down the potential consequences of your chosen responses. How do your reactions to vulnerability affect your wellbeing, relationships, personal growth and ability to connect with others?

- Take a step back to consider the emotions and beliefs that underlie your responses to vulnerability. Are there any common themes that emerge?

- Identify areas where you would like to make positive changes in how you manage vulnerability. Are there alternative strategies that you would like to develop?

- Set goals for yourself based on your reflections. Consider specific actions or practices that can help you embrace vulnerability, cultivate resilience and foster more authentic connections. These could be seeking support from others, practicing self-compassion, engaging in therapy or coaching, or exploring creative outlets.

Remember, this is an activity for non-judgmental personal exploration. The goal is to deepen your self-awareness so you can make conscious choices that support your wellbeing and growth.

5. The power of gratitude

Gratitude is strongly associated with positive emotions, life satisfaction and happiness (Wood et al., 2010). Teachers who cultivate a mindset of gratitude can shift their focus towards the positive aspects of their work, fostering a sense of contentment.

Research suggests that gratitude interventions can decrease symptoms of depression, anxiety and stress (Kerr et al., 2015). For teachers who often face high levels of job-related stress, incorporating gratitude into daily life can help protect mental wellbeing. Gratitude contributes to resilience, enabling us to effectively navigate challenges. Research has established that a positive outlook enables us to bounce back from difficulties (Waters, 2011).

Expressions of gratitude foster positive interpersonal relationships and strengthen social connections. Teachers who demonstrate gratitude to colleagues, students and parents create a supportive and collaborative school community (Froh et al., 2009). Gratitude practices encourage reflection. Keeping a gratitude journal and expressing gratitude through writing can serve as a form of self-care, helping teachers prioritise their wellbeing and manage stress more effectively (Sansone and Sansone, 2010).

OVER TO YOU

Take a moment to write a gratitude letter to someone at work.

- Set aside time to reflect on someone whom you appreciate. Consider a colleague, supervisor or any individual who has made a positive impact on you or the work environment.

- Begin your gratitude letter by addressing the person by their name.

- Express your gratitude sincerely. Share what you appreciate about the person, and specify how their qualities or actions have made a difference.

- Provide concrete examples whenever possible. This helps the recipient understands the impact they've had and makes the letter more meaningful.

- Write from the heart and let your emotions shine through. Be authentic in expressing your gratitude and the feelings it evokes within you.

- Once you've completed your letter, read it through to ensure that it conveys your gratitude effectively. Edit and revise as needed to make it clear and heartfelt.

- Consider the best way to deliver the letter. You can personally hand it to the recipient, leave it on their desk or workspace, or send it electronically.

- If you present the letter in person, express your appreciation verbally. Let the person know that you value their contributions and that their efforts have not gone unnoticed.

- Allow them time and space to absorb your gratitude. They may choose to respond immediately or to reflect on your words privately.

This activity fosters a culture of appreciation and strengthens workplace relationships. It allows you to acknowledge and celebrate the positive impact we have on one another, contributing to a more supportive and cohesive work environment.

6. Stress management

Teachers need to regulate our nervous system so we can better navigate the highs and lows we experience. While there are many risk factors that contribute to our feelings of stress, we need to consider what protective factors we have available to us. Helping keep us resilient, protective factors can include physical health, social supports, personal attributes, problem-solving and cultural identity.

The first thing we should ask ourselves is whether we're taking our MEDS: practicing meditation, exercising, maintaining a healthy diet and getting enough sleep (Burchard, 2020). By incorporating the MEDS approach into our lifestyle, we can strengthen our physical and mental wellbeing. These practices act as protective factors by reducing the impact of stress, enhancing coping mechanisms, and promoting a balanced and resilient response to challenges.

When we are physically healthy, we give ourselves a better chance of being mentally healthy. The more often you make healthy choices about your mind, exercise, diet and sleep, the better your physical and mental health will be.

Meditation

Research suggests that regular meditation can reduce stress levels and improve overall wellbeing. Meditation promotes relaxation, decreases physiological arousal and cultivates mindfulness, which helps us better cope with stressors (Goyal et al., 2014).

Exercise

Regular physical exercise has been shown to have significant stress-reducing effects. Exercise increases endorphin release, improves mood, promotes better sleep and provides a healthy outlet for stress and tension (Salmon, 2001).

Diet

A diet full of nutrient-rich fruits, vegetables, whole grains, lean proteins and healthy fats provides essential nutrients that support brain function and regulate stress hormones (Huang et al., 2019).

Sleep

Sufficient and quality sleep is vital for stress management. Restorative sleep helps regulate stress hormones, enhances cognitive functioning and improves emotional resilience (Grandner et al., 2016).

OVER TO YOU

It's time to track your MEDS.

- Create four columns, one for each component of MEDS.

- In the meditation section, record the number of minutes or sessions you dedicate to meditation each day. You can also note any techniques or apps you use.

- In the exercise section, track the type of exercise you engage in, the duration and the intensity level.

- In the diet section, make a note of the meals, snacks and beverages you consume throughout the day. You can also include any specific dietary goals such as increasing water intake or reducing processed foods.

- In the sleep section, document the hours of sleep you get each night. Note any factors that may influence the quality of your sleep.

- Set a daily reminder or allocate a specific time each day to update your MEDS Tracker. This helps maintain consistency and accountability.

- Regularly review your MEDS Tracker, ideally on a weekly basis. Reflect on patterns and trends, noting areas where you may need to make adjustments.

- Use your tracker to set goals and track progress. For example, you can aim to gradually increase your meditation time, engage in a certain number of weekly workouts, or incorporate more fruits and vegetables into your diet.

- Celebrate milestones and successes along the way. Acknowledge your efforts and the positive impact of MEDS practices on your wellbeing.

You have now created a tangible tool that allows you to visualise your progress. Tracking your MEDS can be motivating, enabling you to cultivate a healthier lifestyle and effectively manage stress.

Final thoughts

As we've learned, thriving requires action on many fronts. Wellbeing is multidimensional, and we need to allow ourselves the time to understand it if we are to make changes in our lives. Our ability to thrive is influenced by the decisions we make, the people we work with and the systemic processes we navigate. Although we will encounter challenges across these areas, we can develop resources that protect and strengthen our wellbeing.

What happens when we prioritise people before performance, connection before content? People feel seen, heard and valued. Collaboration becomes organic because people feel safe enough to share. A culture of genuine diversity is encouraged, and the meaning of our work is enhanced by a common vision. Our mental and physical health improve. Life becomes filled with possibility.

My key message is that you are a human *being*, not a human *doing*. You are encouraged to celebrate yourself, and you deserve to prioritise self-care. Remember, big changes consist of small steps. Your journey to thriving starts today.

Acknowledgements

This book was in my head for years before I had the courage to put pen to paper. With so many ideas and so much to say, I didn't know where to start. Through endless encouragement (some might call nagging) from friends and family, it has become a reality.

I would like to thank all the incredible educators I have worked with over the past 25 years and continue to work with today. From inspiring mentors to people who have challenged me, you all shape the passion and commitment that drives me daily. A special shout-out goes to those schools and principals who have allowed me to conduct professional learning with their staff over the past eight years. I am forever grateful for the opportunities you have given me to learn and grow as I developed my models and programs. It is an honour to serve you and I couldn't do what I do without you.

Thank you also to the inspiring circle of colleagues who I continue to learn from, and who propped up my confidence when I felt lost and alone on the journey. Thanks to Dr Suzy Green for your unwavering faith in me, Dr William DeJean for your ongoing encouragement to shine, Caroline Mansfield for your belief in my writing ability and Dr Christina Curry for your persistence in reminding me to 'just start writing.'

Then of course there is the generous thought-leadership business community that kept me accountable with check-ins to assess writing blocks, weekly word counts and help to scaffold chapters into a sensible order. Thank you to Tracey Ezzard, Selena Fisk and so many others who reminded me that the goal is progress and not perfection. Not to mention Lynne Cazaly, my incredible business coach and mentor. Lynne, you always know the right thing to say when helping me manage my inner critic and remember what is important.

A big shout has to go to my team who help me behind the scenes with their talents and skills. I would be lost without you. Thank you Rosie, Lena, Mark and Maria

Let's not forget the many big-hearted humans doing important work to support teacher wellbeing. Eight years ago, before teacher wellbeing was even a topic, I connected with educators Meg Durham, Katrina Bourke and Ellen Ronalds Keene. Passionate about growing our businesses to support teachers, we have supported each other ever since. Thank you for being my cheer squad and being on the path with me.

Thank you to the team at Amba Press, especially Alicia Cohen. Without you this book wouldn't have happened. After many setbacks and roadblocks in the writing process, you have made the editing and publishing a dream. Thank you for making it all possible.

Thank you also to my crazy family and wonderful friendship groups. There's nothing like family and friends to help you distinguish between what's important and what you need to let go of. Thank you for always cheering me on regardless of the new ideas I come up with. Thank you for your unconditional support, your love and, of course, for making me laugh.

Finally, to my greatest supporter in life, my wife Belinda Carline. Thank you for always believing in me, even when I didn't believe in myself. Thank you for listening to my whinging and complaining as I justified my procrastination. Thank you for asking questions when I presented new ideas to help me strengthen my understanding of what I was saying. Thank you for always being the voice of reason and compassion. You continue to help me become a better human.

The final thanks go to you, the reader. Thank you for your time. I know how precious it is. I sincerely hope that this book has helped you nourish your wellbeing so you can thrive. It's not about being perfect, but about being human. Thank you for all that you do.

You matter.

About the author

Known as the 'keep-it-real' teacher, Daniela Falecki is the founder of Teacher Wellbeing. Having worked for 25 years in Australian schools, Daniela knows the highs and lows of education. She specialises in positive psychology and coaching psychology to merge theory with practical strategies that can be used by busy teachers.

Daniela has a Master of Education (Leadership), a Bachelor of Education (HPE) and a Certificate in Rudolf Steiner Education. She is a licensed Mental Toughness practitioner, certified life coach and neuro-linguistic programming practitioner, and a member of the International Coaching Federation and International Positive Psychology Association.

Daniela has lectured at the University of Sydney and at Western Sydney University, where she was named lecturer of the year in 2014. She has worked in outdoor education as the NSW Senior Manager for the Outdoor Education Group and has consulted on wellbeing for a number of organisations.

Daniela's vision is to see schools change their wellbeing strategies from tokenistic to transformative. She aims to empower teachers with psychological skills and resources to support their own wellbeing, so they can authentically model and teach wellbeing to their students.

Connect with Daniela

Let's keep this conversation going!

Website: teacher-wellbeing.com.au.

Email: daniela@teacher-wellbeing.com.au

LinkedIn: Daniela Falecki

References

Abel, M. H. (2019). *Teachers and burnout: Stress processes in the teaching profession.* Routledge.

Abidin, R. R. (2012). Parenting stress and its impact on parent-child interaction within the family. *Journal of Clinical Child & Adolescent Psychology, 41*(2), 150–160.

Acton, R., & Glasgow, P. (2015). Teacher Wellbeing in Neoliberal Contexts: A Review of the Literature. *Australian Journal of Teacher Education, 40*(8), 6. https://doi.org/10.14221/ajte.2015v40n8.6

Adams, M. (2016). *Coaching Psychology in Schools: Enhancing Performance, Development and Wellbeing.* Routledge/Taylor Francis Group.

Ahghar, G. (2008). The role of school organisational climate in occupational stress among secondary school teachers in Tehran. *International Journal of Occupational Medicine and Environmental Health, 21*(4), 319–329.

Australian Institute for Teaching and School Leadership (AITSL). Improving Teacher Professional Learning. https://www.aitsl.edu.au/teach/improve-practice/improving-teacher-professional-learning

Allen, K. (2016). Roots of Coaching Psychology. *Theory, Research, and Practical Guidelines for Family Life Coaching.* Springer International Publishing. https://doi.org/10.1007/978-3-319-29331-8_2

Algoe, S. B., Haidt, J., & Gable, S. L. (2008). Beyond Reciprocity: Gratitude and Relationships in Everyday Life. *Emotion, 8*(3), 425–429. https://doi.org/10.1037/1528-3542.8.3.425

Arch, J. J., & Craske, M. G. (2006). Mechanisms of mindfulness: Emotion regulation following a focused breathing induction. *Behaviour Research and Therapy, 44*(12), 1849–1858. https://doi.org/10.1016/j.brat.2005.12.007

Avey, J. B., Reichard, R. J., Luthans, F., & Mhatre, K. H. (2011). Meta-analysis of the impact of positive psychological capital on employee attitudes, behaviours, and performance. *Human Resource Development Quarterly, 22*(2), 127–152. https://doi.org/10.1002/hrdq.20070

Bakker, A. B., Hakanen, J. J., Demerouti, E., & Xanthopoulou, D. (2007). Job resources boost work engagement, particularly when job demands are high. *Journal of Educational Psychology, 99*(2), 274–284. https://doi.org/10.1037/0022-0663.99.2.274

Bandura, A. (1997). *Self-Efficacy: The Exercise of Control.* Worth Publishers.

Baumrind, D. (1967). Child care practices anteceding three patterns of preschool behavior. *Genetic Psychology Monographs, 75*(1), 43–88.

Beck, J. S. (2011). *Cognitive Behaviour Therapy: Basics and Beyond* (2nd ed.). Guilford Press.

Berking, M., & Whitley, B. (2014). *Affect Regulation Training: A Practitioners' Manual.* Springer. https://doi.org/10.1007/978-1-4939-1022-9

Bergren, M. D. (2021). Post-COVID-19: Trauma-Informed Care for the School Community. *The Journal of School Nursing, 37*(3), 145. https://doi.org/10.1177/10598405211004709

Bhandari, D. R. (1998). *Plato's Concept of Justice: An Analysis.* Twentieth World Congress of Philosophy, Boston, United States of America. https://www.bu.edu/wcp/Papers/Anci/AnciBhan.htm

Biswas-Diener, R. (2020). The practice of positive psychology coaching. *The Journal of Positive Psychology, 15*(5), 701–704. https://doi.org/10.1080/17439760.2020.1789705

Boyatzis, R. E., & Akrivou, K. (2006). The ideal self as the driver of intentional change. *Journal of Management Development, 25*(7), 624–642. https://doi.org/10.1108/02621710610678454

Brackett, M. A., Bailey, C. S., Hoffman, J. M., & Schoenfeld, T. J. (2019). The ABCs of SEL: A policy guide. Aspen Institute.

Brackett, M. A., & Katulak, N. A. (2007). Emotional Intelligence in the Classroom: Skill-Based Training for Teachers and Students. In J. Ciarrochi & J. D. Mayers (Eds.), *Applying Emotional Intelligence: A Practitioner's Guide* (pp. 1–27). Psychology Press.

Brackett, M. A., Palomera, R., Mojsa-Kaja, J., Reyes, M. R., & Salovey, P. (2019). Emotion skills training for teachers-in-training and in-service professionals: Results from two pilot studies. *Frontiers in Psychology*, 10, 1414.

Brackett, M. A., Rivers, S. E., & Salovey, P. (2011). Emotional intelligence: Implications for personal, social, academic, and workplace success. *Social and Personality Psychology Compass, 5*(1), 88–103. https://doi.org/10.1111/j.1751-9004.2010.00334.x

Brown, B. (2010). *The Gifts of Imperfection: Let Go of Who You Think You're Supposed to Be and Embrace Who You Are.* Hazelden Publishing.

Brown, B. (2016). *Daring Greatly: How the Courage to Be Vulnerable Transforms the Way We Live, Love, Parent, and Lead.* Penguin Life.

Brown, B. (2017). *Rising Strong: How the Ability to Reset Transforms the Way We Live, Love, Parent, and Lead.* Random House.

Brown, B. (2012). *The Power of Vulnerability.* Sounds True.

Bryant, F. B., & Veroff, J. (2007). *Savoring: A new model of positive experience.* Lawrence Erlbaum Associates.

Bryant, F. B., Smart, C. M., & King, S. P. (2005). Using the Past to Enhance the Present: Boosting Happiness Through Positive Reminiscence. *Journal of Happiness Studies, 6*, 227–260. https://doi.org/10.1007/s10902-005-3889-4

Bryk, A. S., & Schneider, B. (2003). Trust in Schools: A Core Resource for School Reform. *Educational Leadership, 60*(6), 40–44.

Burchard, B. (2020). How to Regain Lost Motivation. Brendan Burchard. https://brendon.com/blog/regain-lost-motivation/#:~:text=What%20does%20MEDS%20stand%20for,%2C%20exercise%2C%20diet%2C%20sleep

Cansoy, Ramazan & Parlar, Hanifi & Türko lu, Muhammet. (2020). A Predictor of Teachers' Psychological Well-Being: Teacher Self-Efficacy. *International Online Journal of Educational Sciences, 12*(4), 41–55. https://doi.org/10.15345/iojes.2020.04.003

Caprara, G. V., Barbaranelli, C., Steca, P., & Malone, P. S. (2006). Teachers' self-efficacy beliefs as determinants of job satisfaction and students' academic achievement: A study at the school level. *Journal of School Psychology, 44*(6), 473–490. https://doi.org/10.1016/j.jsp.2006.09.001

Carmeli, A., & Spreitzer, G. (2011). Trust, Connectivity, and Thriving: Implications for Innovative Behaviors at Work. *Journal of Creative Behavior, 43*(3), 169–191. https://doi.org/10.1002/j.2162-6057.2009.tb01313.x

Carson, J. W., Carson, K. M., Gil, K. M., & Baucom, D. H. (2004). Mindfulness-based relationship enhancement. *Behavior Therapy, 35*(3), 471–494. https://doi.org/10.1016/S0005-7894(04)80028-5

CASEL (n.d.). Fundamentals of SEL. https://casel.org/fundamentals-of-sel/

Chew, E. (2017). Enlivening Thriving: Examining Thriving at Work and at Home, Over Time and Across Outcomes. [Doctoral thesis, The University of Western Australia]. The University of Western Australia. https://doi.org/10.26182/5ca6aed621fc3

Chrousos, G. P. (2009). Stress and disorders of the stress system. *Nature Reviews Endocrinology, 5*, 374–381. https://doi.org/10.1038/nrendo.2009.106

Clifton, D. O., & Harter, J. K. (2003). Strengths Investment. In Cameron, K. S., Dutton, J. E., & Quinn, R. E. (Eds.), *Positive Organisational Scholarship* (pp. 111–121). Berrett-Koehler Publishers.

Cohen, S. (2004). Social relationships and health. *American Psychologist, 59*(8), 676–684. https://doi.org/10.1037/0003-066X.59.8.676

Corbu, A., Peláez, Z. MJ., & Salanova, M. (2021). Positive Psychology Micro-Coaching Intervention: Effects on Psychological Capital and Goal-Related Self-Efficacy. *Frontiers in Psychology, 12*, 1–14. https://doi.org/10.3389/fpsyg.2021.566293

Curby, T. W., Rimm-Kaufman, S. E., & Abry, T. (2013). Do emotional support and classroom organisation earlier in the year set the stage for higher quality instruction? *Journal of School Psychology, 51*(5), 557–569. https://doi.org/10.1016/j.jsp.2013.06.001

Curry, J. R. P., & O'Brien, E. R. P. (2012). Shifting to a Wellness Paradigm in Teacher Education: A Promising Practice for Fostering Teacher Stress Reduction, Burnout Resilience, and Promoting Retention. *Ethical Human Psychology and Psychiatry, 14*(3), 178-191. https://doi.org/10.1891/1559-4343.14.3.178

Day, C., & Gu, Q. (2013). *Resilient teachers, resilient schools: Building and sustaining quality in testing times* (1st ed.). Routledge. https://doi.org/10.4324/9780203578490

de Jonge, J., van den Berg, T. I., & Tummers, L. (2018). The impact of perceived emotional job demands on burnout among Dutch teachers: A longitudinal test of the job demands-resources model. *Work & Stress, 32*(2), 146–164.

Deci, E., & Ryan, R. (1990). A motivational Approach to Self: Integration in Personality. In R. Dienstbier (Ed.), *Nebraska Symposium on Motivation, 38*, 237–288.

Deci, E. L., & Ryan, R. M. (2008). Hedonia, eudaimonia, and well-being: an introduction. Journal of Happiness Studies, 9(1), 1-11. https://doi.org/10.1007/s10902-006-9018-1

Diener, E. (1984). Subjective well-being. *Psychological Bulletin, 95*(3), 542–575. https://doi.org/10.1037/0033-2909.95.3.542

Diener, E. (1994). Assessing subjective well-being: Progress and opportunities. *Social Indicators Research, 31*(2), 103–157. https://doi.org/10.1007/BF01207052

Diener, E., Emmons, R.A., Larsen, R.J., & Griffin, S. (1985). The Satisfaction With Life Scale. *Journal of Personality Assessment, 49*(1), 71–75. https://doi.org/10.1207/s15327752jpa4901_13

Diener, E., Wirtz, D., Tov, W., Kim-Prieto, C., Choi, D. W., Oishi, S., & Biswas-Diener, R. (2010). New Well-being Measures: Short Scales to Assess Flourishing and Positive and Negative Feelings. *Social Indicators Research, 97*(2), 143–156. https://doi.org/10.1007/s11205-009-9493-y

Diener, E., & Seligman, M. E. (2002). Very Happy People. *Psychological Science, 13*(1), 81–84. https://doi.org/10.1111/1467-9280.00415

Dodge, R., Daly, A. P., Huyton, J., & Sanders, L. D. (2012). The challenge of defining wellbeing. *International Journal of Wellbeing, 2*(3), 222–235. https://doi.org/10.5502/ijw.v2i3.4

Donohoo, J. (2017). Collective Teacher Efficacy: The Effect Size Research and Six Enabling Conditions. *TheLearningExchange.* https://www.jennidonohoo.com/post/collective-teacher-efficacy-the-effect-size-research-and-six-enabling-conditions

Dryden, W., & Neenan, M. (2020). *Rational Emotive Behaviour Therapy: 100 Key Points and Techniques* (3rd ed.). Routledge.

Durlak, J. A., & Weissberg, R. P. (2005). *A major meta-analysis of positive youth development programs* [Paper presentation]. Presentation at the Annual Meeting of the American Psychological Association, Washington, DC, United States of America.

Durlak, J. A., Weissberg, R. P., Dymnicki, A. B., Taylor, R. D., & Schellinger, K. B. (2011). The Impact of Enhancing Students' Social and Emotional Learning: A Meta-Analysis of School-Based Universal Interventions. *Child Development, 82*(1), 405–432. https://doi.org/10.1111/j.1467-8624.2010.01564.x

Dutton, J. (2003). Fostering high quality connections through respectful engagement. *Stanford Social Innovation Review*, 54–57.

Dutton, J. E., & Heaphy, E. D. (2003). The Power of High-Quality Connections. In K. Cameron & J. Dutton (Eds.), *Positive Organisational Scholarship: Foundations of a New Discipline* (pp. 262–278). Berrett-Koehler Publishers.

Dutton, J. E., Workman, K. M., & Hardin, A. E. (2014). Compassion at Work. *The Annual Review of Organisational Psychology and Organisational Behavior, 1*, 277–304. https://doi.org/10.1146/annurev-orgpsych-031413-091221

Duxbury, L. E., & Higgins, C. (2008). *Work-life balance in Australia in the new millennium: Rhetoric versus reality.*

Dweck, C. S. (2006). *Mindset: The New Psychology of Success.* Random House.

Edmondson, A. (1999). Psychological Safety and Learning Behavior in Work Teams. *Administrative Science Quarterly, 44*(2), 350-383. https://doi.org/10.2307/2666999

Savill-Smith, C., & Scanlan, D. (2022). *Teacher Wellbeing Index 2022.* Education Support UK. https://www.educationsupport.org.uk/media/zoga2r13/teacher-wellbeing-index-2022.pdf

Eisenberg, N., Huerta, & Edwards, A. (2011). Relations of Empathy-Related Responding to Children's and Adolescent's Social Competence. In J. Decetey (Ed.), *Empathy: From Bench to Bedside.* MIT Press Academic. https://doi.org/10.7551/mitpress/9780262016612.003.0009

Elfering, A., Gerhardt, C., Grebner, S., & Muller, U. (2015). Exploring teacher burnout: Differences between primary and secondary teachers. *European Journal of Psychological Assessment, 31*(3), 196–202.

Ellis, A. (1957). Rational psychotherapy and individual psychology. *Journal of Individual Psychology, 13*, 38–44.

Ellis, A. (1994) *Reason and Emotion in Psychotherapy: Comprehensive Method of Treating Human Disturbances.* Stuart (Lyle) Inc.

Emmons, R. A., & McCullough, M. E. (2003). Counting Blessings Versus Burdens: An Experimental Investigation of Gratitude and Subjective Well-Being in Daily Life. *Journal of Personality and Social Psychology, 84*(2), 377–389. https://doi.org/10.1037/0022-3514.84.2.377

Flett, G. L., & Hewitt, P. L. (2002). *Perfectionism and maladjustment: An overview of theoretical, definitional, and treatment issues.* In P. L. Hewitt, & G. L. Flett (Eds.), Perfectionism: Theory, Research, and Treatment (pp. 5–31). American Psychological Association. https://doi.org/10.1037/10458-001

Flook, L., Goldberg, S. B., Pinger, L., Bonus, K., & Davidson, R. J. (2013). Mindfulness for Teachers: A Pilot Study to Assess Effects on Stress, Burnout, and Teaching Efficacy. *Mind, Brain, and Education, 7*(3), 182–195. https://doi.org/10.1111/mbe.12026

Fogg, BJ (2019). *Tiny Habits: The small changes that change everything.* Houghton Mifflin Harcourt

Franco, C., Mañas, I., Cangas, A. J, Moreno, E., Gallego, J. (2010). Reducing teachers' psychological distress through a mindfulness training program. *Spanish Journal of Psychology, 13*(2), 655–666. https://doi.org/10.1017/s1138741600002328

Fredrickson, B. L. (1998). What Good Are Positive Emotions? *Review of General Psychology, 2*(3), 300–319. https://doi.org/10.1037/1089-2680.2.3.300

Fredrickson, B. L., Cohn, M. A., Coffey, K. A., Pek, J., & Finkel, S. M. (2008). Open Hearts Build Lives: Positive Emotions, Induced Through Loving-Kindness Meditation, Build Consequential Personal Resources. *Journal of Personality and Social Psychology, 95*(5), 1045–1062. https://doi.org/10.1037/a0013262

Fredrickson, B. L.; Losada, M. F. (2005). Positive Affect and Complex Dynamics of Human Flourishing. *American Psychologist, 60*(7): 678–686. https://doi.org/10.1037/0003-066x.60.7.678

Friedman, I. A. (2000). Burnout in teachers: Shattered dreams of impeccable professional performance. *Journal of Clinical Psychology, 56*(5), 595–606. https://doi.org/10.1002/(SICI)1097-4679(200005)56:5<595::AID-JCLP2>3.0.CO;2-Q

Froh, J. J., Sefick, W. J., & Emmons, R. A. (2008). Counting blessings in early adolescents: An experimental study of gratitude and subjective well-being. *Journal of School Psychology, 46*(2), 213–233. https://doi.org/10.1016/j.jsp.2007.03.005

Fullan, M., & Hargreaves, A. (2012, June 5). *Reviving Teaching With 'Professional Capital.'* Education Week. https://www.edweek.org/policy-politics/opinion-reviving-teaching-with-professional-capital/2012/06

Galantino, M. L., Baime, M., Maguire, M., Szapary, P. O., & Farrar, J. T. (2005). Association of psychological and physiological measures of stress in health-care professionals during an 8-week mindfulness meditation program: Mindfulness in practice. Stress and Health, 21(4), 255–261. https://doi.org/10.1002/smi.1062

Garner, P.W., Bender, S. L., & Fedor, M. (2018). Mindfulness-based SEL programming to increase preservice teachers' mindfulness and emotional competence. *Psychology in the Schools, 55*(4), 377–390. https://doi.org/10.1002/pits.22114

Ghaith, G., & Yaghi, H. (1997). Relationships among experience, teacher efficacy, and attitudes toward the implementation of instructional innovation. *Teaching and Teacher Education, 13*(4), 451–458. https://doi.org/10.1016/S0742-051X(96)00045-5

Gibson, C., Thomason, B., Margolis, J., Groves, K., Gibson, S., & Franczak, J. (2023). Dignity Inherent and Earned: The Experience of Dignity at Work. Academy of Management Annals, 17(1). https://doi.org/10.5465/annals.2021.0059

Goleman, D. (1995). *Emotional Intelligence: Why It Can Matter More Than IQ.* Bantam Books.

Grandner, M. A., Patel, N. P., Gehrman, P. R., Xie, D., Sha, D., Weaver, T., & Gooneratne, N. (2016). Who Gets the Best Sleep? Ethnic and Socioeconomic Factors Related to Sleep Complaints. *Sleep Medicine, 11*(5), 470--78. https://doi.org/10.1016/j.sleep.2009.10.006

Grant, A. M., & Atad, O. I. (2020). Coaching psychology interventions vs. positive psychology interventions: The measurable benefits of a coaching relationship. *The Journal of Positive Psychology, 17*(4), 532–544. https://doi.org/10.1080/17439760.2021.1871944

Grant, A. M., & Spence, G. B. (2009). Using Coaching and Positive Psychology to Promote a Flourishing Workforce: A Model of Goal-Striving and Mental Health. In Garcea, N., Harrington, S., & Linley, A. P. (Eds.), *Oxford Handbook of Positive Psychology and Work* (pp. 175–188). Oxford University Press. https://doi.org/10.1093/oxfordhb/9780195335446.013.0014

Grant, A. M., Studholme, I., Verma, R., Kirkwood, L., Paton, B., & O'Connor, S. (2017). The impact of leadership coaching in an Australian healthcare setting. *Journal of Health Organisation and Management, 31*(2), 237–252. https://doi.org/10.1108/JHOM-09-2016-0187

Grant, H., & Dweck, C. S. (2003). Clarifying Achievement Goals and Their Impact. *Journal of Personality and Social Psychology, 85*(3), 541–553. https://doi.org/10.1037/0022-3514.85.3.541

Greenberg, M. T., Brown, J. L., & Abenavoli, R. M. (2016). *Teacher stress and health: Effects on teachers, students, and schools.* Pennsylvania State University. https://prevention.psu.edu/publication/teacher-stress-and-health-effects-on-teachers-students-and-schools/

Grossman, P., Niemann, L., Schmidt, S., & Walach, H. (2004). Mindfulness-based stress reduction and health benefits: A meta-analysis. *Journal of Psychosomatic Research, 57*(1), 35–43. https://doi.org/10.1016/S0022-3999(03)00573-7

Guo, J., Marsh, H. W., Parker, P. D., Morin, A. J. S., & Dicke, T. (2017). Extending expectancy-value theory predictions of achievement and aspirations in science: Dimensional comparison processes and expectancy-by-value interactions. Learning and Instruction, 49, 81–91. https://doi.org/10.1016/j.learninstruc.2016.12.007

Hanley, A. W., Garland, E. L., & Fredrickson, B. L. (2020). How does savoring work? Exploring mindfulness and other mechanisms of positive emotion regulation. *Emotion, 20*(7), 1134–1145.

Harris, R. (2017). Sushi Train Metaphor by Dr. Russ Harris. Dr Russ Harris – Acceptance Commitment Therapy. YouTube. https://www.youtube.com/watch?v=tzUoXJVI0wo

Hattie, J. (2009). *Visible learning: A synthesis of over 800 meta-analyses relating to achievement.* Routledge.

Hattie, J. (2012). *Visible Learning for Teachers: Maximizing Impact on Learning.* Routledge.

Hargreaves, A., & Fullan, M. (2012). *Professional Capital: Transforming Teaching in Every School.* Teachers College Press.

Harzer, C., & Ruch, W. (2017). The Role of Character Strengths for Task Performance, Job Dedication, Interpersonal Facilitation, and Organisational Support. *Human Performance, 27*(3), 183–205. https://doi.org/10.1080/08959285.2014.913592

Hascher, T. and Waber, J. (2021). Teacher well-being: A systematic review of the research literature from the year 2000–2019. *Educational Research Review, 34.* https://doi.org/10.1016/j.edurev.2021.100411

Hauge, K. (2019). Teachers' collective professional development in school: A review study, *Cogent Education, 6:1*, 1619223, https://doi.org/10.1080/2331186X.2019.1619223

Hofmann, S. G., Asnaani, A., Vonk, I. J. J., Sawyer, A. T., & Fang, A. (2012). The Efficacy of Cognitive Behavioral Therapy: A Review of Meta-Analyses. *Cognitive Therapy and Research, 36*(5), 427-440. https://doi.org/10.1007/s10608-012-9476-1

Hofmann, S. G., Asnaani, A., Vonk, I. J., Sawyer, A. T., & Fang, A. (2012). The Efficacy of Cognitive Behavioral Therapy: A Review of Meta-Analyses. *Cognitive Therapy and Research, 36*, 427-440. https://doi.org/10.1007/s10608-012-9476-1

Hölzel, B. K., Lazar, S. W., Gard, T., Schuman-Olivier, Z., Vago, D. R., & Ott, U. (2011). How Does Mindfulness Meditation Work? Proposing Mechanisms of Action From a Conceptual and Neural Perspective. *Perspectives on Psychological Science, 6*(6), 537-559. https://doi.org/10.1177/1745691611419671

Hong, S. and Ho, H.-Z. (2005). Direct and Indirect Longitudinal Effects of Parental Involvement on Student Achievement: Second-Order Latent Growth Modeling Across Ethnic Groups. *Journal of Educational Psychology, 97*, 32-42.

Hong, Y.-y., Chiu, C.-y., Dweck, C. S., Lin, D. M.-S., & Wan, W. (1999). Implicit theories, attributions, and coping: A meaning system approach. *Journal of Personality and Social Psychology, 77*(3), 588-599. https://doi.org/10.1037/0022-3514.77.3.588

Howard, S., & Johnson, B. (2004). Resilient teachers: resisting stress and burnout. *Social Psychology of Education: An International Journal, 7*(4), 399-420. https://doi.org/10.1007/s11218-004-0975-0

Hurley, D. B., Kwon, P., & Koo, M. (2016). Optimism, savoring, and coping with a stressor: A study of freshmen adjusting to university life. *Journal of Happiness Studies, 17*(3), 1181-1191.

Huta, V., & Waterman, A. S. (2014). Eudaimonia and Its Distinction from Hedonia: Developing a Classification and Terminology for Understanding Conceptual and Operational Definitions. *Journal of Happiness Studies, 15*(6), 1425-1456. https://doi.org/10.1007/s10902-013-9485-0

Hwang, Y.-S., Bartlett, B., Greben, M., & Hand, K. (2017). A systematic review of mindfulness interventions for in-service teachers: A tool to enhance teacher wellbeing and performance. *Teaching and Teacher Education, 64*, 26-42. https://doi.org/10.1016/j.tate.2017.01.015

International Organization for Standardization (2018). ISO 45001—Occupational Health and Safety. https://www.iso.org/iso-45001-occupational-health-and-safety.html

International Organization for Standardization (2021). ISO 45003—Occupational Health and Safety. https://www.iso.org/iso-45003-occupational-health-and-safety.html

Jacka, F. N., Pasco, J. A., Mykletun, A., Williams, L. J., Hodge, A. M., O'Reilly, S. L., Nicholson, G. C., Kotowicz, M. A., & Berk, M. (2010). Association of Western and traditional diets with depression and anxiety in women. *American Journal of Psychiatry, 167*(3), 305-311. https://doi.org/10.1176/appi.ajp.2009.09060881

Jennings, P. A. (2015). *Mindfulness for teachers: Simple Skills for Peace and Productivity in the Classroom* (1st ed.). W. W. Norton and Company.

Jennings, P. A., & Greenberg, M. T. (2009). The Prosocial Classroom: Teacher Social and Emotional Competence in Relation to Student and Classroom Outcomes. *Review of Educational Research, 79*(1), 491-525. https://doi.org/10.3102/0034654308325693

Jennings, P. A., Brown, J. L., Frank, J. L., Doyle, S., Oh, Y., Davis, R., Rasheed, D., DeWeese, A., Demauro, A. A., Cham, H., & Greenberg, M. T. (2017). Impacts of the CARE for Teachers program on teachers' social and emotional competence and classroom interactions. *Journal of Educational Psychology, 109*(7), 1010–1028. https://doi.org/10.1037/edu0000187

Jennings, P. A., Frank, J. L., Snowberg, K. E., Coccia, M. A., & Greenberg, M. T. (2013). Improving classroom learning environments by Cultivating Awareness and Resilience in Education (CARE): Results of a randomized controlled trial. *School Psychology Quarterly, 28*(4), 374–390. https://doi.org/10.1037/spq0000035

Jiang, G., Jiang, L., & Liu, D. (2015). The relationship between psychological capital and teachers' job burnout: A meta-analysis. *Social Indicators Research, 122*(3), 731–747.

Johnson, B., Down, B., Le Cornu, R., Peters, J., Sullivan, A., Pearce, J., & Hunter, J. (2014). Promoting early career teacher resilience: a framework for understanding and acting. *Teachers and Teaching, 20*(5), 530-546. https://doi.org/10.1080/13540602.2014.937957

Jones, S. M., & Bouffard, S. M. (2012). Social and Emotional Learning in Schools: From Programs to Strategies and Commentaries. *Social Policy Report, 26*(4), 1–33. https://doi.org/10.1002/j.2379-3988.2012.tb00073.x

Jorm, A. F., Morgan, A. J., Hetrick, S. E. (2008). Relaxation for Depression. *Cochrane Database of Systematic Reviews, 4.* https://doi.org/10.1002/14651858.CD007142.pub2

Kabat-Zinn, J. (2005). *Coming to Our Senses: Healing Ourselves and the World Through Mindfulness.* Hachette UK.

Kaihoi, C. A., Bottiani, J. H., & Bradshaw, C. P. (2022). Teachers Supporting Teachers: A Social Network Perspective on Collegial Stress Support and Emotional Wellbeing Among Elementary and Middle School Educators. *School Mental Health, 14*(4), 1070–1085. https://doi.org/10.1007/s12310-022-09529-y

Kaplan, A., et al. (2018). Examining the role of teachers› personal life stories in teaching: Opportunities and dilemmas. *Teachers and Teaching, 24*(8), 861–878.

Kaplan, S., Cortina, J., Ruark, G. A., & Ozer, E. (2019). Organisational reputation: A review. *Journal of Management, 45*(6), 2475–2505.

Kerr, S. L., O'Donovan, A., & Pepping, C. A. (2015). Can gratitude and kindness interventions enhance well-being in a clinical sample? *Journal of Happiness Studies: An Interdisciplinary Forum on Subjective Well-being, 16*(1), 17–36. https://doi.org/10.1007/s10902-013-9492-1

Kemeny, M. E., Foltz, C., Cavanagh, J. F., Cullen, M., Giese-Davis, J., Jennings, P., Rosenberg, E. L., Gillath, O. Shaver, P. R., Wallace, B. A., & Ekman, P. (2012). Contemplative/emotion training reduces negative emotional behavior and promotes prosocial responses. *Emotion, 12*(2), 338–350. https://doi.org/10.1037/a0026118

Keng, S. L., Smoski, M. J., & Robins, C. J. (2011). Effects of Mindfulness on Psychological Health: A Review of Empirical Studies. *Clinical Psychology Review, 31*(6), 1041–1056. https://doi.org/10.1016/j.cpr.2011.04.006

Keyes, C. L. M. (2006). Subjective Well-being in Mental Health and Human Development Research Worldwide: An Introduction. *Social Indicators Research, 77*(1), 1–10. https://doi.org/10.1007/s11205-005-5550-3

Keyes, C. L. M. (2014). Mental Health as a Complete State: How the Salutogenic Perspective Completes the Picture. In: Bauer G.F. & Hämmig, O (Eds.), Bridging Occupational, Organisational and Public Health (pp. 179–92). Springer. https://doi.org/10.1007/978-94-007-5640-3_11

Kiken, L. G., Garland, E. L., Bluth, K., Palsson, O. S., & Gaylord, S. A. (2015). From a state to a trait: Trajectories of state mindfulness in meditation during intervention predict changes in trait mindfulness. *Personality and Individual Differences, 81*, 41–46. https://doi.org/10.1016/j.paid.2014.12.044

Kim, J., Youngs, P., & Frank, K. (2017). Burnout contagion: Is it due to early career teachers' social networks or organisational exposure? *Teaching and Teacher Education, 66*, 250–260. https://doi.org/10.1016/j.tate.2017.04.017

King, L. A. (2001). The Health Benefits of Writing about Life Goals. *Society for Personality and Social Psychology Bulletin, 27*(7), 798-807. https://doi.org/10.1177/0146167201277003

Kjell, O. N. E. (2011). Sustainable Well-Being: A Potential Synergy Between Sustainability and Well-Being Research. *Review of General Psychology, 15*(3), 255–266. https://doi.org/10.1037/a0024603

Klassen, R. M., & Tze, V. M. C. (2014). Teachers' self-efficacy, personality, and teaching effectiveness: A meta-analysis. *Educational Research Review, 12*, 59–76. https://doi.org/10.1016/j.edurev.2014.06.001

Kleine, A-K., Rudolph, C. W., & Zacher, H. (2019). Thriving at work: A meta-analysis. *Journal of Organisational Behavior, 40*(9–10), 973–999. https://doi.org/10.1002/job.2375

Kubzansky LD, Huffman J, Boehm J, Hernandez R, et al. (2018) Positive Psychological Well-Being and Cardiovascular Disease: JACC Health Promotion Series. *Journal of the American College of Cardiology, 72*(12), 1382–1396. https://doi.org/10.1016/j.jacc.2018.07.042

Klusmann, U., Kunter, M., Trautwein, U., Lüdtke, O., & Baumert, J. (2008). Teachers' occupational well-being and quality of instruction: The important role of self-regulatory patterns. *Journal of Educational Psychology, 100*(3), 702–715. https://doi.org/10.1037/0022-0663.100.3.702

Kristeller, J. L., & Wolever, R. Q. (2010). Mindfulness-based eating awareness training for treating binge eating disorder: The conceptual foundation. *Eating Disorders, 19*(1), 49–61. https://doi.org/10.1080/10640266.2011.533605

Kyriacou, C., & Chien, P. Y. (2004). Teacher stress in Taiwanese primary schools. *Journal of Educational Enquiry, 5*(2), 86–104.

Kyriacou, C., & Kunc, R. (2007). Beginning teachers' expectations of teaching. *Teaching and Teacher Education, 23*(8), 1246-1257. https://doi.org/10.1016/j.tate.2006.06.002

Leahy, K. E., Birch, L. L., & Rolls, B. J. (2008). Reducing the energy density of multiple meals decreases perceived appetite and increases feelings of satiety in children. *The American Journal of Clinical Nutrition, 88*(6), 1459–1468. https://doi.org/10.3945/ajcn.2008.26522

Leeds Beckett University. (2018, January 23). *Pupil progress held back by teachers' poor mental health.* https://www.leedsbeckett.ac.uk/news/0118-mental-health-survey/

Li, S., Zhang, B., Yang, X., Hu, H., Zhang, X., & Wang, Q. (2021). The prevalence and risk factors of psychological impacts on Chinese health care workers during COVID-19 outbreak: A cross-sectional survey study. *Frontiers in Psychology, 12*, 597134.

Locke, E. A., & Latham, G. P. (Eds.). (2013). *New developments in goal-setting and task-performance.* Routledge. https://doi.org/10.4324/9780203082744

Lomas, J., Stough, C., Hansen, K., & Downey, L. A. (2012). Brief report: Emotional intelligence, victimisation and bullying in adolescents. *Journal of Adolescence, 35*(1), 207–211. https://doi.org/10.1016/j.adolescence.2011.03.002

Lopez, S. J., & Louis, M. C. (2009). The Principles of Strengths-Based Education. *Journal of College and Character, 10*(4), 1–7. https://doi.org/10.2202/1940-1639.1041

Luthans, F., & Youssef, C. M. (2007). Emerging Positive Organisational Behavior. *Journal of Management, 33*(3), 321–349. https://doi.org/10.1177/0149206307300814

Luthans, F., Avey, J. B., Avolio, B. J., & Peterson, S. (2010). The Development and Resulting Performance Impact of Positive Psychological Capital. *Human Resource Development Quarterly, 21*(1), 41–67. https://doi.org/10.1002/hrdq.20034

Luthans, F., Youssef, C. M., & Avolio, B. J. (2007). *Psychological Capital: Developing the Human Competitive Edge*. Oxford University Press. https://doi.org/10.1093/acprof:oso/9780195187526.001.0001

Luthans, F., Vogelgesang, G. R., & Lester, P. B. (2006). Developing the Psychological Capital of Resiliency. *Human Resource Development Review, 5*(1), 25–44. https://doi.org/10.1177/1534484305285335

Lyubomirsky, S., & Layous, K. (2013). How Do Simple Positive Activities Increase Well-Being? *Current Directions in Psychological Science, 22*(1), 57–62. https://doi.org/10.1177/0963721412469809

Mansfield, C., Beltman, S., Broadley, T., Weatherby-Fell, N. & MacNish, D. (2015). *BriTE: Building Resilience in Teacher Education*. Murdoch University. https://resiliencetoolkit.org.uk/wp-content/uploads/2018/01/brite-building-resilience-in-teacher-education.pdf

Mansfield, C. F., Beltman, S., Price, A., & McConney, A. (2012). "Don't sweat the small stuff": Understanding teacher resilience at the chalkface. *Teaching and Teacher Education, 28*(3), 357–367. https://doi.org/10.1016/j.tate.2011.11.001

Marais-Opperman, V., Rothmann, S. I., & van Eeden, C. (2021). Stress, flourishing and intention to leave of teachers: Does coping type matter? *SA Journal of Industrial Psychology, 47*(1), 1–11. https://dx.doi.org/10.4102/sajip.v47i0.1834

Maslach, C., & Leiter, M. P. (2016). Understanding the burnout experience: recent research and its implications for psychiatry. *World Psychiatry, 15*(2), 103–111. https://doi.org/10.1002/wps.20311

Maslow, A. H. (1943). A theory of human motivation. *Psychological Review, 50*(4), 370–396. https://doi.org/10.1037/h0054346

Mayer, J. D., & Salovey, P. (1997). What is Emotional Intelligence? In P. Salovey & D. Sluyter (Eds.), *Emotional Development and Emotional Intelligence: Educational Implications for Educators* (pp. 3–34). Basic Books.

McCallum, F., & Price, D. (2010). Well teachers, well students. *The Journal of Student Wellbeing, 4*(1), 19–34. https://doi.org/10.21913/JSW.v4i1.599

McCallum, F., & Price, D. (2012). Keeping teacher wellbeing on the agenda. Professional Educator, 11(2), 4–7.

McGonigal, K. (2016). *The Upside of Stress: Why Stress Is Good for You, and How to Get Good at It*. Avery.

McGonigal, K. (2019). *The Joy of Movement: How Exercise Helps Us Find Happiness, Hope, Connection, and Courage*. Avery.

McMahon, D. M., Wernsing, T. S., & Luthans, F. (2011). The role of creative problem-solving in resilience and job satisfaction: A longitudinal study. *Journal of Occupational and Organisational Psychology, 84*(1), 153–173.

MentalHelp.net (n.d.). Cognitive Restructuring. https://www.mentalhelp.net/stress/cognitive-restructuring/

Nagoski, A. & Nagoski, E. (2019). *Burnout: The secret to unlocking the stress cycle.* Ballantine Books.

Neff, K. D. (2003). Self-Compassion: An Alternative Conceptualisation of a Healthy Attitude Toward Oneself. *Self and Identity, 2*(2), 85–101. https://doi.org/10.1080/15298860309032

NeiTA (2021). *ACE Teachers Report Card 2021: Teachers' perceptions of education and their profession.* Australian College of Educators. https://www.austcolled.com.au/wp-content/uploads/2021/10/NEiTA-ACE-Teachers-Report-Card-2021.pdf

Nelson, S. K. & Lyubomirsky, S. (2012). Finding happiness: Tailoring positive activities for optimal well-being benefits. M. Tugade, M. Shiota, & L. Kirby (Eds.), *Handbook of positive emotions.* New York: Guilford.

Newport, C. (2016). *Deep Work: Rules for focused success in a distracted world.* Grand Central Publishing.

Niemiec, R. M. (2017). *Character Strengths Interventions: A Field Guide for Practitioners.* Hogrefe Publishing.

Nolen-Hoeksema, S., & Morrow, J. (1993). Effects of rumination and distraction on naturally occurring depressed mood. *Cognition and Emotion, 7*(6), 561–570. https://doi.org/10.1080/02699939308409206

O'Brien, K., & Mosco, J. (2012). Positive Parent-Child Relationships. In Roffey, S. (Ed.), *Positive Relationships: Evidence Based Practice across the World* (pp. 91–107). Springer. https://doi.org/10.1007/978-94-007-2147-0_6

OECD. (2013). *OECD guidelines on Measuring Subjective Well-Being,* OECD Publishing. https://doi.org/10.1787/9789264191655-en.

Özbilgin, M. F., Tatli, A., Karatas-Ozkan, M., & Kunday, Ö. (2016). Social support for high-potential women in management and professional occupations: Insights from an emerging economy. *Human Resource Management, 55*(2), 327–347.

Palmer, S., & Whybrow, A. (2007). *Handbook of Coaching Psychology: A Guide for Practitioners.* Routledge.

Park, N., Peterson, C., & Seligman, M. E. P. (2006). Character strengths in fifty-four nations and the fifty US states. *The Journal of Positive Psychology, 1*(3), 118–129. https://doi.org/10.1080/17439760600619567

Parker, P. D., Martin, A. J., Colmar, S., & Liem, G. A. (2012). Teachers' workplace well-being: Exploring a process model of goal orientation, coping behavior, engagement, and burnout. *Teaching and Teacher Education, 28*(4), 503–513. http://dx.doi.org/10.1016/j.tate.2012.01.001

Pavot, W. G., Diener, E., Colvin, C. R., & Sandvik, E. (1991). Further validation of the Satisfaction with Life Scale: Evidence for the cross-method convergence of well-being measures. *Journal of Personality Assessment, 57*(1), 149–161. https://doi.org/10.1207/s15327752jpa5701_17

Pianta, R. C., Belsky, J., Vandergrift, N., Houts, R., & Morrison, F. (2008). Classroom Effects on Children's Achievement Trajectories in Elementary School. *American Educational Research Journal, 45*(2), 365–397. https://doi.org/10.3102/0002831207308230

Pianta, R. C., Hamre, B. K., & Allen, J. P. (2012). Teacher–student relationships and engagement: Conceptualizing, measuring, and improving the capacity of classroom interactions. In S. L. Christenson, A. L., Reschly, & C. Wylie (Eds.), *Handbook of Research on Student Engagement* (pp. 365–386). Springer Science + Business Media. https://doi.org/10.1007/978-1-4614-2018-7_17

Pilcher, J. J., & Walters, A. S. (1997). How sleep deprivation affects psychological variables related to college students' cognitive performance. *Journal of American College Health, 46*(3), 121–126. https://doi.org/10.1080/07448489709595597

Plutchik, R. (1980). *Emotions: A Psychoevolutionary Synthesis.* Harper & Row.

Porath, C. (2022). *Mastering Community: The Surprising Ways Coming Together Moves Us from Surviving to Thriving.* Little Brown.

Porath, C. L., Gibson, C. B., & Spreitzer, G. M. (2022). To thrive or not to thrive: Pathways for sustaining thriving at work. *Research in Organisational Behavior, 42.* https://doi.org/10.1016/j.riob.2023.100185

Porath, C., Spreitzer, G., Gibson, C., Garnett, F.G. (2012). Thriving at work: Toward its measurement, construct validation, and theoretical refinement. *Journal of Organisational Behavior, 33*(2), 250–275. https://doi.org/10.1002/job.756

PWC Australia. (2014). *Creating a mentally healthy workplace: Return on investment analysis.* https://www.pwc.com.au/publications/pdf/beyondblue-workplace-roi-may14.pdf

Quoidbach, J., Berry, E. V., Hansenne, M., & Mikolajczak, M. (2010). Positive emotion regulation and well-being: Comparing the impact of eight savoring and dampening strategies. *Personality and Individual Differences, 49*(5), 368–373. https://doi.org/10.1016/j.paid.2010.03.048

Rath, T., & Conchie, B. (2008). *Strengths Based Leadership: Great Leaders, Teams, and Why People Follow.* Gallup Press.

Raver, C. C., Jones, S. M., Li-Grining, C. P., Metzger, M., Champion, K. M., & Sardin, L. (2008). Improving Preschool Classroom Processes: Preliminary Findings from a Randomized Trial Implemented in Head Start Settings. *Early Childhood Research Quarterly, 23*(1), 10–26. https://doi.org/10.1016/j.ecresq.2007.09.001

Redelinghuys, J.J., Rothmann, S., & Botha, E. (2019). Workplace flourishing: Measurement, antecedents and outcomes. *SA Journal of Industrial Psychology, 45.* https://doi.org/10.4102/sajip.v45i0.1549

Richards, J. (2012). Teacher Stress and Coping Strategies: A National Snapshot. *The Educational Forum, 76*(3), 299-316. https://doi.org/10.1080/00131725.2012.682837

Richardson, G. E. (2002). The metatheory of resilience and resiliency. *Journal of Clinical Psychology, 58*(3), 307–321. https://doi.org/10.1002/jclp.10020

Richardson, P. W., Watt, H. M., & Devos, C. (2013). Types of professional and emotional coping among beginning teachers. *Emotion and School: Understanding How the Hidden Curriculum Influences Relationships, Leadership, Teaching, and Learning* (pp. 229–253). Emerald Group Publishing Limited. https://doi.org/10.1108/S1479-3687(2013)0000018018

Roeser, R. W., Schonert-Reichl, K. A., Jha, A., Cullen, M., Wallace, L., Wilensky, R., Oberle, E., Thomson, K., Taylor, C., & Harrison, J. (2013). Mindfulness training and reductions in teacher stress and burnout: Results from two randomized, waitlist-control field trials. *Journal of Educational Psychology, 105*(3), 787–804. https://doi.org/10.1037/a0032093

Roorda, D. L., Koomen, H. M. Y., Spilt, J. L., & Oort, F. J. (2011). The Influence of Affective Teacher-Student Relationships on Students' School Engagement and Achievement: A Meta-Analytic Approach. *Review of Educational Research, 81*(4), 493–529. https://doi.org/10.3102/0034654311421793

Roffey, S. (2008). Emotional literacy and the ecology of school wellbeing. *Educational and Child Psychology, 25*(2), 29-39.

Ryan, R. M., & Deci, E. L. (2017). *Self-determination theory: Basic psychological needs in motivation, development, and wellness.* The Guilford Press. https://doi.org/10.1521/978.14625/28806

Ryff, C. D. (1989). Happiness is everything, or is it? Explorations on the meaning of psychological well-being. *Journal of Personality and Social Psychology. 57*(6): 1069–1081. https://doi.org/10.1037/0022-3514.57.6.1069

Safe Work Australia (2019). *Model Health and Safety Regulations.* Parliamentary Counsel's Committee. https://www.safeworkaustralia.gov.au/system/files/documents/1902/model-whs-regulations-15-january-2019.pdf

Salmon, P. (2001). Effects of physical exercise on anxiety, depression, and sensitivity to stress: A unifying theory. *Clinical Psychology Review, 21*(1), 33–61. https://doi.org/10.1016/s0272-7358(99)00032-X

Salovey, P., & Mayer, J. D. (1990). Emotional intelligence. *Imagination, Cognition and Personality, 9*(3), 185–211. https://doi.org/10.2190/DUGG-P24E-52WK-6CDG

Salovey, P., Brackett, M. A., & Mayer, J. D. (Eds.). (2004). *Emotional intelligence: Key readings on the Mayer and Salovey model.* Dude Publishing.

Sansone, R. A., & Sansone, L. A. (2010). Gratitude and Well Being: The Benefits of Appreciation. *Psychiatry (Edgmont), 7*(11), 18–22.

Scheuch, K., E. Haufe and R. Seibt (2015). Teachers' Health. *Deutsches Aerzteblatt Online.* http://dx.doi.org/10.3238/arztebl.2015.0347

Schonert-Reichl, K. A., Oberle, E., Lawlor, M. S., Abbott, D., Thomson, K., Oberlander, T. F., & Diamond, A. (2015). Enhancing Cognitive and Social-Emotional Development Through a Simple-to-Administer Mindfulness-Based School Program for Elementary School Children: A Randomized Controlled Trial. *Developmental Psychology, 51*(1), 52–66. https://doi.org/10.1037 per cent2Fa0038454

Schonert-Reichl, K. A., & Lawlor, M. S. (2010). The Effects of a Mindfulness-Based Education Program on Pre-and Early Adolescents' Well-Being and Social and Emotional Competence. *Mindfulness, 1,* 137–151. https://doi.org/10.1007/s12671-010-0011-8

Schuch, F. B., Vancampfort, D., Firth, J., Rosenbaum, S., Ward, P. B., Silva, E. S., Hallgren, M., Leon, A. P. D., Dunn, A. L., Deslandes, A. C., Fleck, M. P., Carvalho, A. F., & Stubbs, B. (2018). Physical Activity and Incident Depression: A Meta-Analysis of Prospective Cohort Studies. *The American Journal of Psychiatry, 175*(7), 631–648. https://doi.org/10.1176/appi.ajp.2018.17111194

Schussler, D. L., Jennings, P. A., Sharp, J. E., & Frank, J. L. (2016). Improving teacher awareness and well-being through CARE: A qualitative analysis of the underlying mechanisms. *Mindfulness, 7*(1), 130–142. https://doi.org/10.1007/s12671-015-0422-7

Schön, D. A. (1987). *Educating the reflective practitioner: Toward a new design for teaching and learning in the professions.* Jossey-Bass.

See, S.M. et al. (2022). The Australian Principal Occupational Health, Safety and Wellbeing Survey 2022 Data. ACU Institute for Positive Psychology & Education. https://www.healthandwellbeing.org/reports/AU/2022_ACU_Principals_HWB_Final_Report.pdf

Seligman, M.E. P. (1972). Learned Helplessness. *Annual Review of Medicine, 23,* 407–412. https://doi.org/10.1146/annurev.me.23.020172.002203

Seligman, M. E. P. (1991). *Learned Optimism.* Knopf.

Seligman, M. E. P. (2011). *Flourish: A visionary new understanding of happiness and well-being.* Free Press.

Seligman, M. E. P, Ernst, R. M., Gillham, J., Reivich, K., & Linkins, M. (2009). Positive education: positive psychology and classroom interventions. *Oxford Review of Education, 35*(3), 293-311. https://doi.org/10.1080/03054980902934563

Seligman, M. E. P., Steen, T. A., Park, N., & Peterson, C. (2005). Positive Psychology Progress: Empirical Validation of Interventions. *American Psychologist, 60*(5), 410-421. https://doi.org/10.1037/0003-066X.60.5.410

Shapiro, S. L., Jazaieri, H., & Goldin, P. R. (2012). Mindfulness-based stress reduction effects on moral reasoning and decision making. *Journal of Positive Psychology, 7*(6), 504-515. https://doi.org/10.1080/17439760.2012.723732

Sheldon, K. M., & Lyubomirsky, S. (2007). How to increase and sustain positive emotion: The effects of expressing gratitude and visualizing best possible selves. *The Journal of Positive Psychology, 1*(2), 73-82. https://doi.org/10.1080/17439760500510676

Shimazu, A., Schaufeli, W. B., Kamiyama, K., & Kawakami, N. (2015). Workaholism vs. work engagement: The two different predictors of future well-being and performance. *International Journal of Behavioral Medicine, 22*(1), 18-23. https://doi.org/10.1007/s12529-014-9410-x

Sinek, S. (2009). *Start with Why: How Great Leaders Inspire Everyone to Take Action*. TEDxPuget Sound. https://www.ted.com/talks/simon_sinek_how_great_leaders_inspire_action

Skaalvik, E. M., & Skaalvik, S. (2010). Teacher self-efficacy and teacher burnout: A study of relations. *Teaching and Teacher Education, 26*(4), 1059-1069. https://doi.org/10.1016/j.tate.2009.11.001

Skaalvik, E. M., & Skaalvik, S. (2011). Teacher job satisfaction and motivation to leave the teaching profession: Relations with school context, feeling of belonging, and emotional exhaustion. *Teaching and Teacher Education, 27*(6), 1029-1038. https://doi.org/10.1016/j.tate.2011.04.001

Skaalvik, E. M., & Skaalvik, S. (2014). Teacher Self-Efficacy and Perceived Autonomy: Relations with Teacher Engagement, Job Satisfaction, and Emotional Exhaustion. *Psychological Reports, 114*(1), 68-77. https://doi.org/10.2466/14.02.PR0.114k14w0

Snyder, C. R. (2002). Hope theory: Rainbows in the mind. Psychological Inquiry, 13(4), 249-275. https://doi.org/10.1207/S15327965PLI1304_01

Spreitzer, G. M., Cameron, L., & Garrett, L. (2017). Alternative work arrangements: Two images of the new world of work. *Annual Review of Organizational Psychology and Organizational Behavior, 4*, 473-499. https://doi.org/10.1146/annurev-orgpsych-032516-113332

Spreitzer, G., Porath, C.L., & Gibson, C.B. (2012). Toward human sustainability: How to enable more thriving at work. *Organisational Dynamics, 41*(2), 155-162. https://doi.org/10.1016/j.orgdyn.2012.01.009

Spreitzer, G., Sutcliffe, K., Dutton, J., Sonenshein, S., & Grant, A. M. (2005). A socially embedded model of thriving at work. *Organisation Science, 16*(5), 537-549. https://doi.org/10.1287/orsc.1050.0153

Steger, M. F. (2012). Experiencing meaning in life: Optimal functioning at the nexus of well-being, psychopathology, and spirituality. In Wong, P. T. P. (Ed.), *The Human Quest for Meaning: Theories, Research, and Applications* (pp. 165-184). Routledge.

Steger, M. F., Kashdan, T. B., & Oishi, S. (2008). Being good by doing good: Daily eudaimonic activity and well-being. *Journal of Research in Personality, 42*(1), 22-42. https://doi.org/10.1016/j.jrp.2007.03.004

Stoeber, J., & Rennert, D. (2008). Perfectionism in school teachers: Relations with stress appraisals, coping styles, and burnout. *Anxiety, Stress, & Coping, 21*(1), 37–53. https://doi.org/10.1080/10615800701742461

Sutton, R. E., & Wheatley, K. F. (2003). Teachers' Emotions and Teaching: A Review of the Literature and Directions for Future Research. *Educational Psychology Review, 15*, 327–358. https://doi.org/10.1023/A:1026131715856

Tang, Y. Y., Ma, Y., Wang, J., Fan, Y., Feng, S., Lu, Q., Yu, Q., Sui, D., Rothbart, M. K., Fan, M., & Posner, M. I. (2007). Short-term meditation training improves attention and self-regulation. *Proceedings of the National Academy of Sciences, 104*(43), 17152–17156. https://doi.org/10.1073/pnas.0707678104

Tschannen-Moran, M., & Hoy, A. W. (2001). Teacher efficacy: capturing an elusive construct. *Teaching and Teacher Education, 17*(7), 783–805. https://doi.org/10.1016/S0742-051X(01)00036-1

Tugade, M. M., & Fredrickson, B. L. (2004). Resilient individuals use positive emotions to bounce back from negative emotional experiences. *Journal of Personality and Social Psychology, 86*(2), 320–333. https://doi.org/10.1037/0022-3514.86.2.320

Viac, C., & Fraser, P. (2020). Teachers' well-being: A framework for data collection and analysis. *OECD Education Working Papers, 213*. https://doi.org/10.1787/c36fc9d3-en

Ware, H., & Kitsantas, A. (2007). Teacher and Collective Efficacy Beliefs as Predictors of Professional Commitment. *Teaching and Teacher Education, 100*(5), 303–310. https://doi.org/10.3200/JOER.100.5.303-310

Waterman, A.S. (2008). Reconsidering happiness: A eudaimonist's perspective. *Journal of Positive Psychology, 3*(4), 234–252. https://doi.org/10.1080/17439760802303002

Waters, L. (2011). A Review of School-Based Positive Psychology Interventions. *The Australian Educational and Developmental Psychologist, 28*(2), 75–90. https://doi.org/10.1375/aedp.28.2.75

Waters, L. E., Loton, D., & Jach, H. K. (2019). Does strength-based parenting predict academic achievement? The mediating effects of perseverance and engagement. *Journal of Happiness Studies: An Interdisciplinary Forum on Subjective Well-Being, 20*(4), 1121–1140. https://doi.org/10.1007/s10902-018-9983-1

Waters, T., Marzano, R. J., & McNulty, B. (2003). *Balanced Leadership: What 30 years of Research Tells us about the Effect of Leadership on Student Achievement.* McREL International.

Watson, A. (n.d.). 40 Hour Teacher Workweek. https://join.40htw.com/

Watson, D., Clark, L.A., & Tellegen, A. (1988). Development and validation of brief measures of positive and negative affect: The PANAS scales. *Journal of Personality and Social Psychology, 54*(6), 1063–1070. https://doi.org/10.1037//0022-3514.54.6.1063

Wentzel, K. R. (2012). Teacher-Student Relationships and Adolescent Competence at School. In T. Wubbels, P. D. Brok, J. V. Tartwijk & J. Levy (Eds.), *Interpersonal Relations in Education: An Overview of Contemporary Research* (pp. 19–36). Springer. https://doi.org/10.1007/978-94-6091-939-8_2

Wood, A. M., Froh, J. J., & Geraghty, A. W. (2010). Gratitude and well-being: a review and theoretical integration. *Clinical Psychology Review, 30*(7), 890-905. https://doi.org/10.1016/j.cpr.2010.03.005

World Health Organization (WHO) (1947). *World Health Organization Act 1947.* https://www.legislation.gov.au/Details/C2016C00962

World Health Organization (WHO) (2005). *Promoting mental health: concepts, emerging evidence, practice.* https://www.who.int/publications/i/item/9241562943

World Health Organization (WHO) (2019/2021). *International Classification of Diseases, Eleventh Revision (ICD-11).* https://icd.who.int/browse11

Xanthopoulou, D., Bakker, A. B., Demerouti, E., & Schaufeli, W. B. (2007). The role of personal resources in the job demands-resources model. *International Journal of Stress Management, 14*(2), 121–141. https://doi.org/10.1037/1072-5245.14.2.121

Yeager, D. S., & Dweck, C. S. (2012). Mindsets That Promote Resilience: When Students Believe that Personal Characteristics Can Be Developed. *Educational Psychologist, 47*(4), 302–314. https://doi.org/10.1080/00461520.2012.722805

www.ingramcontent.com/pod-product-compliance
Lightning Source LLC
Chambersburg PA
CBHW071211210326
41597CB00016B/1766